OTHER TITLES OF INTEREST FROM ST. LUCIE PRESS

Ecosystem Management for Sustainability: Illustrated by an International Biosphere Reserve Cooperative

Ecological Integrity and the Management of Ecosystems

Development, Environment, and Global Dysfunction: Toward Sustainable Recovery

Handbook of Trace Elements in the Environment

Economic Theory for Environmentalists

Resolving Environmental Conflict: Toward Sustainable Community Development

The Everglades: A Threatened Wilderness

Everglades: The Ecosystem and Its Restoration

The Everglades Handbook: Understanding the Ecosystem

Ecology and Management of Tidal Marshes: A Model from the Gulf of Mexico

Environmental Effects of Mining

Environmental Fate and Effects of Pulp and Paper Mill Effluents

Sustainable Forestry: Philosophy, Science and Economics

Contamination of Groundwater

Naturally-Occurring Radioactive Material: Principles and Practices

Chemistry and Health Effects of Hazardous Chemicals

Trash to Cash: How Businesses Can Save Money and Increase Profits

For more information about these titles call, fax or write:

St. Lucie Press
100 E. Linton Blvd., Suite 403B
Delray Beach, FL 33483
TEL (561) 274-9906 ● FAX (561) 274-9927

ECOLOGY
and the
BIOSPHERE
Principles and Problems

Sharon La Bonde Hanks

S^t_L

St. Lucie Press
Delray Beach, Florida

S^{t}L

Published by
St. Lucie Press
100 E. Linton Blvd., Suite 403B
Delray Beach, FL 33483

To my family and friends
for their continuous support and encouragement

ACKNOWLEDGMENTS

I would like to thank all of my colleagues and students through the years. They have been a constant source of knowledge and growth for me, both personally and professionally. A special thanks to Paul Cirielli, whose computer expertise made my computer questions and technical problems go away. I would also like to thank Lynn Kloss, my editor, without whose help this book would not have been possible.

CONTENTS

LIST OF TABLES AND FIGURES

Tables

Figures

ACKNOWLEDGMENTS

The author wishes to thank the following copyright holders for permission to reprint their work.

Wadsworth Publishing Company for the following material from *Living in the* Environment by G. Tyler Miller, 4th ed., © 1985: Figure 2-27: Monthly Average Carbon Dioxide Concentrations from 1960-1990; Table 2-18: World Water Resources and Table 2-19: Average U.S. Water Use in 1983.

Worldwatch Institute for the following figures and tables: Figure 2-7: Annual Increase in World Population, 1950-1993; Figure 2-9: Annual Oyster Catch on the Chesapeake Bay; Figure 2-15: United States Corn Yields per Hectare, 1866-1993; Table 2-8: World Grain Production and Fertilizer Use, 1950-1993; Figure 2-16: World Grain Output Per Person, 1950-1993, with Projections to 2030; Table 2-9: Per Capita Grain Use and Consumption of Livestock Products in Selected Countries, 1990; Figure 2-23: World Grain Use, Total, and for Feed, 1960-1992; Table 2-13: Worldwide Land Degradation Due to Human Activity, 1945-1994; Figure 2-28: World Paper Consumption, 1931-1991; Figure 2-29: U.S. Materials Consumption and Population, 1900-1991; Table 2-20: Sources of Marine Pollution; Figure 2-34: World Fertilizer Use 1950-1994; Table 2-23: Energy Used in the Production and Recycling of Selected Building Materials in the United Kingdom.

G. Tyler Miller. Every effort has been made to contact G. Tyler Miller for permission to reprint Table 2-28: Examples of Introduced Species and the Damage They Cause. The author would be very interested to hear from the copyright owner.

Every effort has been made to contact the original source for Figure 2-17: Energy Subsidies for Food Production by Food Type. The author would be very interested to hear from the copyright owner.

True understanding in an individual has two attributes: awareness and action.

Who can enjoy enlightenment and remain indifferent to the suffering in the world?

Only those who increase their service along with their understanding can be called men and women of the way.

Source: Brian Walker. *Hua Hu Ching: The Unknown Teachings of Lao Tzu*. 1994, Harper Collins, revised edition.

FOREWORD

For the past several years, I have been actively involved in teaching a non-majors biology course on the biosphere, and each semester, I ask students what they want to learn. The responses are consistent: Students want to know how science relates to their lives, how the biosphere works, what is currently wrong with the biosphere (current problems), and what they can do to make a difference. This book is my response to these expressed desires. The primary goals of this text are to give the reader a basic understanding of science, of the complexities of the biosphere, of how it works, and of how this information relates to the reader's life.

Many of the texts that are currently available present a large quantity of mathematical or statistical information. Because many students taking this course are, by definition, not scientists, and actually do not think that they can understand science, they do not need to be bombarded with statistics that do not increase their understanding of the material. In contrast, this text has been written for non-science majors, with relevant terminology explained and illustrated, and with assignments that help students relate to and work with the material.

Because this book was designed for a one-semester course, it is relatively short, and it focuses on the material that every person — not just every scientist — needs in order to make intelligent choices about the future of the earth. It presents specific scientific concepts integrated with current biospheric problems, so that students can easily make the necessary connections to their own lives.

I have been teaching college students since 1965 in a variety of science courses. Some of these courses were for students majoring in biology, and some were for students who were non-science majors. Most colleges require that students take one or more science courses. These non-major science courses are usually one semester in length and have no prerequisites. A wide variety of courses has been developed specifically for such students. Depending on the institutions at which they are offered, these non-major courses have been given different designations including general education, liberal studies, core requirements, and distribution requirements.

I have found that students who take non-major science courses do so for a variety of reasons, ranging from those who have an interest in science to those who are taking the courses only because they are required to do so. These students are sometimes rather hostile or anxious about the subject. Not only do student attitudes vary, but the amount of scientific background that students bring with them can range from one semester to three years of science courses. Students who come to the class with little science background, and those who feel that the course is a waste of their time, must see the importance of this material to their lives. To engage students with the material is, thus, one of my current interests, and I hope to achieve this by having students interact with the material in a variety of ways. I have designed interactive assignments for the end of each chapter.

I have used the following format for each topic:

- an explanation of basic scientific theories, laws, principles, concepts, and processes
- a discussion of how the scientific theories, laws, principles, concepts, and processes relate to the biosphere
- a discussion of current biosphere problems related to the scientific theories, laws, principles, concepts, and processes
- a discussion of the effect of human activity on the biosphere
- a discussion of possible solutions to the biospheric problems
- writing activities to engage the reader with the material.

This approach links the basic scientific information to students' everyday lives, helping them make connections.

The assignments at the end of each section are structured so that students will not only learn content, but also will increase their academic abilities and cognitive development. By completing the assignments, they will learn to:

- explain basic ecological principles
- broaden their perspective
- use their imagination
- critically evaluate the impact of human activities
- explain the structure and function of an ecosystem
- relate biology to other parts of their lives.

The following is a partial list of the current biospheric problems that are discussed in this text: overpopulation, global warming, nutrient cycle disruptions, soil loss, productivity (food), loss of species diversity, water pollution, air pollution, loss of tropical rainforests and temperate forests, ozone depletion, toxic waste, and acid rain.

After reading the text and completing the assignments, students should be able to analyze current environmental and/or ecological issues. When they read an article from a secondary source, such as a newspaper or magazine, or see a news report on television concerning a biospheric or environmental issue, they should be able to:

- clearly state the issue, question, or problem under discussion
- identify which scientific laws, theories, principles, concepts, or processes are important to understanding the matter under discussion
- formulate a series of questions that need to be addressed when discussing the issue, question, or problem
- construct arguments for continuing current practices
- construct arguments for changing or discontinuing current practices
- decide which solution they prefer and, in a well-developed discussion of one or two paragraphs, explain their reasoning.

Since this may be the only science course that students take, I hope to give them the background, understanding, and interest in science that will help them relate to science throughout their lives.

INTRODUCTION

Students often ask why they need to know anything about science, because it does not relate to *their* lives. Science is not something that impacts on only a few individuals in the world; science impacts on *you* and on everyone, every day. This book deals with the biosphere (the world) and some of its current problems, and points out how these issues relate to your life. It is divided into three areas: Part 1 deals with what science is and what you need to know about it. Part 2 deals with the biosphere (the world), how it works, and its current problems. Part 3 deals with what you can do about these problems.

I have structured this book as a series of questions, because I feel that this is a useful way to learn. By the time you finish this book, you should be able to ask necessary questions and make decisions about current biospheric problems and their proposed solutions. (See Appendix I: Approach to Biospheric Issues.)

You probably have opinions about everything. So do most people. Does that mean that all opinions are equal? No. Some opinions are based on facts. These carry more weight than do those based only on what someone feels. You need information to form a reasonable opinion. This book will begin to help you gain some of the information that you need to form a fact-based opinion, and it will point out where you need to obtain more information. It is perfectly all right to not have an opinion about something, but it is important to know when you need to gather more information before you form an opinion or make a decision. By the time you finish this book, you should know what questions you need to ask and what kinds of information you need to gather.

Each of you brings to this book your own experiences, knowledge, perceptions, and biases about science, college, and life. Your experiences color what you hear, see, and read, and they will shape how you interact with the material in this text. I use the word *experience* because this book is not a lecture. For you to get the most out of this book, you must work with the material and experience it.

Some of you have already had science classes and know what to expect from a science text. Others are in a science course for the first time. Some of you are anxious about science and mathematics, and this will also color how you work with this book. It doesn't matter which group of students you fall into, because this book is written for the non-science student. It is designed to be approachable and to make learning science a painless process. There will, by necessity, be some mathematics in the book, but I have kept the math to the minimum you need to get a clear picture of the topic under discussion. An understanding of dwindling animal or plant populations, for example, requires that you know something about how, and why, these numbers are calculated.

You should review what you know and what you believe about each topic before you read each chapter. This process will help you sort out what is new and different in the chapter. Some of your previous knowledge may be incorrect, and some of your beliefs may differ from those presented here. If you do not know where you stand, it will be difficult to identify where your misconceptions are and where you may want to change your opinions. At the end of this introduction, I will explain a quick and easy way for you to identify your knowledge and preconceptions.

Each of you has a preferred learning style, and therefore, each of you will respond differently to different teaching styles. If there is a match between your learning style and the instructor's teaching style, everyone is at ease. However, a mismatch between learning and teaching styles can result in anger, frustration, anxiety, and hostility for all parties. Learning styles range from hearing (listening), reading (seeing), writing, speaking (teaching others or working in small groups), hands-on experience, and drawing (imaging or model building).

Just as there are different styles of learning, there are different ways of processing information (e.g., reading a text, making a diagram, or writing notes). The more ways you process information, the better is your chance of remembering it. Learning is *not* just memorizing facts and being able to repeat them. You learn by being involved (doing) and by working with the information so that you can transform it into your own experience. In this way, you can consciously relate new information to your existing knowledge or preconceptions. You make connections and distinctions when you transform information into your own words. Learning requires that you work with the material or transform the information.

Transformation can occur through writing, speaking, or drawing. Writing allows you to convert what you have read, seen, or heard into your own words. Drawing (graphic representation, illustration, or imaging) allows you to transform the information into a visual form, for example, a graph. Imaging puts the information in long-term memory and gives you a different perspective. The results of imaging are greater understanding, aided learning, and better recall. At the end of each topic, I have included assignments that will allow you to work with the information so you can get the most out of the book.

Writing is not only a way of thinking. It is also a manifestation of the thought process. As you write, you transform new information, manipulating it to fit the task, making new connections and combinations, and evaluating and ordering it. Writing makes you reflect on and analyze information. Through writing, you can discover what you think and gain new insight into yourself, and the material. You can "own" the information by expressing it in your own words. Writing is a tool of learning. It is a visible account to which you can refer.

The assignments at the end of each chapter give you the opportunity to work with the material while practicing and improving all of your academic skills and abilities. As you do the assignments, you will learn the material in a different way than you do by reading or listening to a lecture. The more ways you work with the information, the easier it will be for you to incorporate it into your thinking and memory.

Many students complain that science is hard to learn because of all the vocabulary. It is true that science does have terminology that is unfamiliar to non-scientists, but any time you take a course in a new discipline, like sociology, philosophy, or communications, you have to learn the vocabulary of that field. Learning science vocabulary is like learning a foreign language. The more you use it, and the more ways in which you use it, the easier it will be to incorporate it into your knowledge.

In this book, I have given definitions for each scientific term the first time it is used. A glossary is provided at the end of the book for your reference. I have kept the vocabulary to a minimum, but it is important that you are familiar with some terms because, as you read articles or see programs about science throughout your life, these are the words that will be used. If you don't understand the terminology, you will not understand what is being discussed.

This book is concerned with the structure and function of the biosphere. The biosphere is the layer around the earth that contains all life. It is a global ecosystem that includes all living organisms (biotic) and nonliving material (abiotic). The biosphere can be thought of as a spider's web. It is a beautiful, complex system in which all the parts interrelate and interconnect. Like a spider's web, if you pull one strand, the entire web vibrates and changes shape.

If you break one strand, the entire web is affected. If you break enough strands, the web collapses and ceases to function.

The same principles are true about the biosphere, but, unlike a spider's web, the biosphere cannot be rewoven overnight. Because of the interrelatedness and interconnectedness of the biosphere, it is difficult to discuss in an ordered way. I will, however, write about one thing at a time and use examples to clarify the information. When I choose an example to make a point, I will refer to the other parts of the biosphere that affect the example, and I will refer to the same example in several places in the text.

Biospheric problems are not the result of only one thing, and they do not impact on the biosphere in only one way. For example, the problem of deforestation results not only in the loss of the forest system, but it also affects the soil, the water, the species living in the area, the weather, and nutrient cycling. By the end of this book, you will understand that the biosphere functions not as a system composed of separate, easily defined, and limited parts, but as a complex, interactive whole.

Before you continue with this book, I want to share with you a technique that will assist you in identifying your knowledge, perceptions, and biases about the material. This technique, which you may have used in other courses, is called freewriting. You should use this technique to anchor yourself about the topic before you start each new section or chapter. Freewriting takes about three to five minutes, and it is private. No one has to see what you write except you. Simply write the topic at the top of a piece of paper, then quickly write everything that you associate with that topic. This is a stream-of-consciousness activity. Write all the words or phrases that come to your mind. Do not organize or edit what you write. Write *everything*. Include facts, feelings, names: anything that you associate with the topic or area. When you are finished, look at what you have written and identify your knowledge, feelings, and attitudes about the topic. When you have finished freewriting, keep the paper with you as you read and refer to it, noting what does not agree with your initial knowledge and perceptions, and what confirms what you already know.

This device will help you to identify where you start so that you can continually examine the new material presented, and learn to make reasoned decisions on how you will act toward the earth, both while you are in this course and throughout your life.

PART 1: SCIENCE

WHAT IS SCIENCE?

What is "science"? If you look the word *science* up in the dictionary, you will find several definitions. I have chosen to define science as "*people* seeking to discover facts so that they can gain a *better understanding* of the relationships in the *natural* and *physical* world." The definition sounds simple and straightforward, but like most definitions, the words carry with them a tremendous amount of assumed information, or background knowledge. The words in the definition that I will discuss more fully are: people, relationship, and physical.

I chose to use the word *people*, not scientists, in the definition to point out the importance of remembering that scientists are people first and scientists second. Like you, scientists are unique individuals with different background knowledge, preconceptions, and biases. They bring all this with them to their work (research). Scientists are influenced directly (consciously) and indirectly (unconsciously) by their culture, economic, religious, political, and social backgrounds. Therefore, all of these factors determine how science is done by affecting what questions are asked or investigated, how an experiment is designed, what is chosen for observation, how results are interpreted, and what conclusions and inferences are drawn.

An example of how these factors can influence which questions are asked and how they are asked is the research done on heart attacks in the United States. Most of the research on the cause and treatment of heart attacks has been done on white males. The results of these research projects have then been applied to other populations — white women and people of color. Only recently has research been expanded to include other groups. Why was the research done only on white males? Who has been asking the questions and designing the research experiments? White males. Why is this important? The results of research on one population do not necessarily apply to other populations. The causes and treatments of heart attacks for white males may not be the same as the causes and treatments for white females or people of color.

Another example of how scientists have been influenced by their culture and background is the research done on brain size, human worth, and intelligence. Scientists have done research to determine which group of humans have the largest brain. The assumptions were that large brain size indicated higher intelligence and human worth, and that smaller brain size indicated lower intelligence and, consequently, less human worth. Researchers measured brains and found that the brains of the white males studied were larger than the brains of the women and black males. The problem with this research is that the researchers ignored the already known facts that physical size and age influence the size of the brain. Physically larger people and younger people have larger brains and physically smaller people and older people have smaller brains. The fact that the white males in the studies were physically larger than the black males and women selected would predict that the white males would, by comparison, have larger brains. The researchers, however, ignored the existing information and concluded that white males had larger brains and therefore were smarter or more intelligent, and by extension had more human worth.

These two examples show how scientists, as people, are influenced by their own perspectives, and how that can be reflected in their work. There are many examples of how science has been used to support specific cultural and economic ideas. Does this mean that scientists have intentionally conducted research to achieve specific results? When past research has been reviewed by other scientists, it appears that some scientists have consciously done research to support certain concepts, but that most of the biased or slanted research has been done unconsciously.

The second part of the definition of science, "to gain a *better understanding* of relationships," indicates that scientists don't know everything there is to know about the natural and physical world, and that they are still looking for answers. Relationships are complex interactions, and therefore, understanding them is not a simple process.

Science is often referred to as a "state-of-the-art" discipline. This means that scientists are continually discovering new information about how the natural and physical world works. As new technologies and methodologies are developed, the production of new information is accelerated. What we knew last year or last month may be out of date today. New information requires that our understanding of how things work be revised.

Interesting examples of science as a state-of-the-art discipline can be found in the medical profession. In the early 1900s, based on the available information, doctors prescribed small amounts of arsenic to alleviate the pain of migraine headaches. Today no one would prescribe arsenic for migraines because our understanding has changed. *State of the art* means that we have to make decisions based on what we know today, even though new information may change our understanding tomorrow.

New technologies, like biotechnology and bioengineering, produce new information daily. This means that people can't always rely on old information. Often, new information affects you directly. You may know someone who has had a serious medical problem. One thing that all patients are advised to do is to get a second opinion before they make a decision about how to take care of a medical problem.

Why is this a good idea? Doctors are human beings, and therefore they diagnose a problem and suggest a treatment depending on their background, where they went to medical school, and what their medical interests are. All of these factors can affect a doctor's recommendations. For example, doctors who attended institutions that teach that a specific disease should be treated by surgery may not seriously consider other methods of treatment. Doctors who graduated 10 years ago may not be as current on new information or procedures as those who graduated more recently. On the other hand, newly graduated doctors may not have the experience to see all the underlying factors of a patient's condition. In other words, by going for a second opinion, you increase the amount of information you have when you make a decision. Sometimes it is even necessary to get a third opinion. The more information you obtain, the more informed your decision will be.

So, scientists are people first, and scientists second. Science as a state-of-the-art discipline changes every day, but it is not always possible to wait for new information. Yes, in a month or a year there may be a better understanding of a problem, but in many cases you, and for that matter, the **biosphere**, may not have the time to wait. Waiting may only make the problem worse, and in fact, waiting may allow the problem to deteriorate to the point that nothing can correct it. Just because science may not have all the answers does not mean that people can put off making decisions.

Does this mean that because scientists do not have everything figured out, you should discount what they know? No, you need to understand that information may change about an issue or problem, but that scientists are giving you the best information they have at the time. To argue that scientists do not have all the answers and therefore should be ignored does not address the problem at hand.

The third part of our definition is that scientists want to understand the "natural and physical world." This sets limits on what scientists can study. There is often confusion among nonscientists about what questions can be addressed by science. Science is limited to dealing with those things that can be directly or indirectly sensed. Direct sensing is hearing, seeing, touching, smelling, and tasting. Indirect sensing means those things that can be monitored with instrumentation. Science, by definition, does not deal with things that cannot be sensed. Science cannot define beauty or truth. These concepts are important, but they are not part of what science does. Science is limited to the natural and physical world.

WHO DOES SCIENCE AND WHY? WHAT IS A SCIENTIST?

Scientists are fascinated by the natural and physical world; they want to discover how things work, to explain what is going on. They are generally amazed by the complexity and wonderment of the natural and physical world. If I were to ask you to describe a scientist, you would probably list some of the following characteristics: old, male, white, intelligent, authority figure, wears a white lab coat, wears glasses, dedicated, boring, eccentric, etc. Scientists are often described as objective (noninvolved) and rational (nonemotional). These are common stereotypes of scientists in the United States.

How you think of scientists has been shaped by the media and by your own experience. Like all stereotypes, this one is simplistic. Yes, scientists do try to be objective and rational about their research, but if you look into the history of science, you will find many examples of scientists who have not been completely rational or objective.

Scientists, like all other people, sometimes have difficulty separating their personal beliefs and desires from their research. For example, the race to discover the structure of the basic genetic material in chromosomes, DNA (deoxyribonucleic acid), was an international race. The scientific community knew that whoever found the answer first would be famous. This led some scientists to do unethical things, such as using other researchers' information without permission and without giving credit. Society often believes that scientists are completely objective, are separated from their emotions, and do not make mistakes. These myths may lead people to a false understanding of what science is and what kinds of things they have to pay attention to as they read about science or think about what scientists do.

Scientists bring to their research their individual background, perceptions, and biases. These factors affect the kinds of questions they ask, the way they ask these questions, and how they interpret the results of their experiments. This alteration of perception is not an intentional process, and if you do not look for the individual's approach, you may be misled about their research. Just as others have their "baggage" to bring to science, so do you. You need to know what you bring with you when you read about science. Objectivity does not just happen; you must pay careful attention to the above motivations in yourself and in others.

Science is a synthetic and collaborative discipline. That means that what a scientist does is based on the work of other researchers. An individual may add to the scientific knowledge base, but that contribution is based on what he or she knew about the problem before asking the question. Historically, the person who asks the "key" question or synthesizes knowledge in a new way that increases our understanding of life gets the credit and the glory. But it is also essential that credit be given to those who produced the knowledge base from which a scientist works, and since most scientists do not work alone, it is essential that credit be given to all who helped do the research, whether they were research assistants or professors. Science is collaborative, and those who do the work should get the credit. This is one of the ethical issues that a scientist must face. Scientists have moral and ethical standards to meet as human beings. Science does not excuse a researcher from these responsibilities.

Many new discoveries and ideas are named after a single individual. The structural model of DNA is named after two scientists, Watson and Crick, and the basic understanding of how evolution works is named after Charles Darwin. Even though discoveries are often named for individual scientists, science is not an individual endeavor. Science is a synthesis discipline, built on the current and past work of thousands of scientists. Science is an accumulation of information from many different disciplines (e.g., mathematics, physics, chemistry, geology, biology).

Why then do individuals get credit? Ideas, concepts, theories, and laws are named after the person or persons who take the existing scientific knowledge and, with new information of their own or by seeing a new pattern for existing information, formulate a new understanding of how something works. They synthesize something new out of existing information. Sometimes this synthesis is a straightforward process, and sometimes others cannot understand how a researcher arrived at the new idea. Sometimes the researcher makes what could be described as an intuitive jump — what you might call the "light bulb effect." You don't know exactly how you figured it out, but all of the sudden you have the answer. Remember that the researcher has internalized information from past scientists before that light bulb went on.

WHAT DO SCIENTISTS DO?
HOW DO THEY DO IT?

Scientists want to understand how things work. To do this they ask questions, then try to find the answers to these questions. The problem-solving process used by scientists is referred to as the **scientific method**. Many people think that because the thought process has a name, it is difficult and hard to understand. However, you carry out a similar process many times a day. The scientific method is really just organized common sense. Scientists are trained to use this thought process. Learning to do the scientific method is like learning to drive a car or to read. At first you do each step carefully, paying close attention so that you do it correctly. After you have practiced long enough, you don't think about all the steps required to turn a corner or to read a book, you just do them. Once scientists learn to do the scientific method, they do it automatically.

The scientific method is a continuous process. To discuss the process, I have broken it down into three parts: observation, formulation, and prediction. The first part of the scientific method is **observation**. This is an accurate recording of what a person observes about some aspect of the natural or physical world. Observations can be made directly (people record what they actually see) or indirectly (an instrument indicates or measures what is happening). "Observing" is different from "seeing" in the everyday sense. We "see" all the time but very seldom do we go around "observing." Observing means paying close attention to what is going on. Observing in the scientific method also means recording what is observed. These recordings must be as complete as possible so that the scientist can go back and know exactly what was observed. A recording that says that you saw a bird fly by is not a careful observation. You would not know what kind of bird it was, when it flew by, how fast it flew, etc. The recording of information or observations is often referred to as the collecting or recording of **data**. The plural form of the word is *data; datum* is the singular form. Each observation is a datum and all the observations are the data. After the data have been recorded, the scientist tries to understand how it relates to the question that was asked.

The second part of the scientific method is the **formulation** of a possible explanation of what was observed. This possible explanation is called an **hypothesis**. To develop an hypothesis, it is necessary for the scientist to synthesize all the available information that is related to the observation.

After an hypothesis has been formulated, the scientist progresses to the third part of the thought process, **prediction**. The scientist tries to test the hypothesis by making predictions. In other words, the researcher tests the hypothesis by designing an experiment. Scientists like to design experiments in which they can control all aspects of the experiment (temperature, amount of light, food, space, interactions, etc.) However, controlled experiments are not always ethical, nor are they always the best way to conduct an experiment.

To demonstrate how the scientific method works, I will use the following example: You are fond of apples, therefore you eat lots of apples. Over five years you eat 500 apples, and each time you eat an apple you record its color and taste. At the end of five years, you have 500 observations (data). Of the 500 apples you ate, 300 were red and sweet and 200 were green and tart. You then read all the information you can find on apples. You decide that you are primarily interested in green apples, and you formulate a possible explanation of why green apples are tart. Your hypothesis is that green apples are tart because they contain a specific chemical. We will call the chemical "X." You will then design an experiment to test your hypothetical explanation for the tartness in green apples.

You take 500 green apples and measure whether chemical X is present and whether the apples are tart. If all green apples tested have chemical X and are tart, you would have data to support your hypothesis. However, if all of the apples had chemical X, but some were sweet, you would have to rethink your hypothesis and proceed to design another experiment to test a new hypothesis.

You can see that your experiments may lead to new questions and that the process will continue until you understand the tartness of green apples. If you were to write out this process, it would look like this: *If* all green apples are tart *because* they contain chemical X, *then if* I test 500 green apples for the presence of chemical X and they all contain that chemical and are tart, my hypothesis has some supportive evidence. If you and other scientists repeat this experiment, and over time all the data support your hypothesis, it will gain acceptance in the scientific community.

It is important to understand that even though you have supported your hypothesis, you have not proved it 100%. To prove your hypothesis about the tartness of green apples 100%, you would have to test all of the green apples in the world for the presence of chemical X and tartness.

An hypothesis is never proven 100%. It only has more or less supportive data. This is true for all levels of explanation from hypothesis to theories to laws. There is always the chance that as new information is discovered, and as new technologies are developed, some hypotheses, theories, and laws may be shown to be incorrect. One interesting thing about science is that it does change with new information. When new information is produced, even long-established ideas can be changed. Once again, we see that science is a state-of-the-art discipline.

An example of this change is the way that geologists have revised their theories of how the continents move. In 1935, geologists believed that continents were static (they didn't move over the earth's surface). However, Wegener, based on his interpretation of available information, hypothesized that land masses did move. His hypothesis was not accepted because he did not have sufficient data to persuade other geologists. However, by the 1960s, new methodologies in geologic research produced information that led to new ideas of how continents moved. This information supported Wegener's idea. The science of geology now accepts his theory; it is called the theory of continental drift.

There are two categories of explanatory hypothesis: **causal** and **teleological**. Causal explanations explain an observation by finding out what caused the observed event to happen. Teleological explanations explain an observation by specifying the purpose that underlies the observed action. A teleological explanation implies that the organism knows, has a reason, or has a goal in mind, and therefore does something to achieve it. For example, if you observed Canada geese flying south in the fall, a teleological explanation might be that the geese knew it was warmer in the south in the winter, or that they knew there would be more food available there. Because you can't ask geese if they know it is warmer in Florida in the winter, you cannot test a teleological hypothesis.

A causal explanation would try to discover what events caused the geese to fly south. The causal explanation might be that the shortening of the day length or the decrease in temperature caused physiological changes in the geese, and these changes resulted in their migration. You could then design an experiment to test your causal hypothesis. You could change the day length or temperature, then determine if there were physiological changes in the geese. Causal explanations can be tested, therefore, they are the kind of explanatory hypothesis that scientists use. This is the second restriction on scientific research: an hypothesis must be **testable**.

Once a causal explanatory hypothesis has been formulated, a researcher must decide what kind of information (observations or data) will be collected. Information (data) can be divided into two major categories: quantitative and qualitative. Quantitative information can be expressed in numbers: You can count the number of eggs a hen lays when being fed different diets. Qualitative information does not lend itself to numerical expression: If you were studying the color of the eggs being produced, you might find that the colors range from white to off-white to cream to beige to brown. The problem here is that what you mean by beige or brown might not be the color I think of as beige or brown. Scientists prefer to use quantitative data because it is clearly understood by other researchers. Three eggs means the same thing to both of us. Quantitative data is precise, and other researchers can independently judge the conclusions made by the original researcher.

Once a scientist has decided on the kind of data that will be collected to test an hypothesis, he or she must decide on how many observations will be made or how many individuals will be tested before the data is analyzed and conclusions drawn. Why do scientists collect many observations before they make a conclusion? Why don't they just make a few observations? No two organisms are exactly alike; there is always variation between organisms. You are a human being, but the person next to you is not exactly like you. In the same way, no two chickens, trees, or bacteria are exactly the same. There is always some variation between the organisms being tested.

How does this variation affect the experiment? Let's say that you are interested in the average height of the students in your class. If you randomly choose two students and they are both six feet tall, you might conclude that the average student height is six feet. If you had randomly chosen two students who were five-feet-two inches tall, you would conclude that the average height for students in your class was five-foot-two. Both conclusions would be incorrect because you did not measure enough students to overcome the variation in height in the population of students in your class.

There is no set number of observations for all experiments. The number of observations will depend on the amount of variation in the population being studied. The greater the amount of variation, the more observations and the more data will have to be collected. The researcher wants to make sure that the difference in the data is due to what happened in the experiment, not to the variation in the population. For example, if the scientist wants to test the effect of diets on egg production, enough chickens will have to be tested so that any difference in egg production can be attributed to the diets fed the hens, not on the egg-laying variation between hens.

Because data is recorded as it is observed, it is not in an organized state, and will make little sense. To make data useful, it must first be **organized**. If you were interested in the effect of different diets on hen growth rates, you would collect data for each hen over a period of time, then organize the data so that you could make sense of it. After the data is organized, it can be **analyzed**.

Analyzing data and drawing conclusions are the final steps in the process of doing science. How you organize and analyze your data depends on what question you ask or what you want to know. Once you have decided on your approach to the data (what question you ask), you will analyze it to see if you can find the answer to your question. The analysis may be very simple (e.g., comparing average hen weights) or very complicated, requiring the use of sophisticated statistics and computers.

After the data is analyzed, you will **synthesize** the information from your work with information from other researchers, draw conclusions, and relate all of this information to the original question. When all of these steps have been completed, you are ready to ask another question. Then the process begins again.

HOW DO SCIENTISTS ANALYZE DATA?

Statistical analysis is really a manipulation of data. If two scientists ask different questions or use different methods of analysis (manipulation), they may use the same data but arrive at different results. Often different interpretations of data depend on the perceptions and biases of the researcher. Sometimes these differences may depend on who pays the researcher (vested interests and conflict of interest). Which questions are asked or not asked is also important. This is another reason that it is important to gather information from several sources before you make a decision on any issue.

The interpretation of data depends on the underlying assumptions made by the researcher. An **assumption** is a fact or idea that the researcher accepts as true. Many researchers do not detail their assumptions when they present the results of their research. It is important to clearly understand the assumptions, however, because they influence how the research is conducted and how the data is interpreted. The conclusions and inferences made from a set of data may vary depending on who is looking at the data. These two words, *conclusion* and *inference*, are commonly used by researchers when they discuss their work. They are frequently used interchangeably, although they do not mean the same thing. You must be careful to ask exactly what can be concluded from the research and what the scientist is inferring. A **conclusion** is a decision based on the analysis of the data collected. An **inference** is a more tentative generalization based on this analysis. The degree of certainty is higher for a conclusion than it is for an inference. For example, if you see smoke billowing out of a window, you can conclude that there is smoke, but you cannot conclude that there is a fire in the room. You can *infer* that there is a fire, but the only way to *conclude* that there is a fire is to see the flames. Often it is unclear what is a conclusion and what is an inference. Learning how to read a graph can help you decide for yourself if the conclusions and inferences being drawn are supported by the data.

The numerical results from research are often presented as a graph or visual representation. You should be able to look at a graph and describe what it illustrates. A good graph clearly indicates what the X-axis and Y-axis represent, and the informative title should tell what the graph is about.

Compare Figures 1-1 and 1-2. The two graphs present data from the same experiment on the effects of three different diets on weight gain in hogs. Use the graphs and their titles to draw some conclusions. However, you may not come to the same conclusions about the experiment because of the way the graphs have been constructed. For example, Figure 1-1 gives a time frame of 100 days, while Figure 1-2 gives no time frame. From Figure 1-2, you might conclude that because all three diets yield the same weight gain, diet doesn't make any difference. But if you were to look at weight gain over time (Figure 1-1), you would see that, in fact, hogs do gain weight at different rates with different diets.

Graphic illustrations can lead to different conclusions, depending on how they are constructed. A good graph should be self-explanatory. Anyone should be able to understand what the experiment was about and what the data indicate. A good graphic presentation takes careful planning and thought.

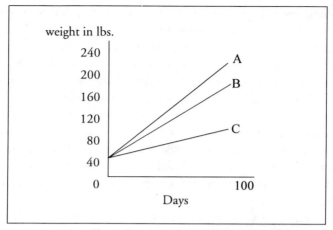

Figure 1-1: The Effects of Three Different Diets on Hog Weight Increase (showing weight gain over a 100-day period)

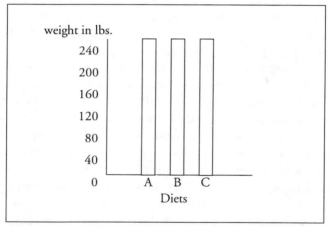

Figure 1-2: The Effects of Three Different Diets on Hog Weight Increase (showing total weight gain in hogs)

Presenting data in a graphic format makes the information easier to understand. Most of the graphs in this book are either **bar graphs** or **point-to-point graphs**. The type of graph used depends on the kind of data being presented. A bar graph is used when the information is discontinuous or when a range of measurements is pictured. In Figure 1-2, for example, diet is a discontinuous characteristic, so it is represented by a bar graph. A point-to-point graph presents data when the information is continuous. In Figure 1-1, for example, the weight gain of the hogs over a 100-day period is a continuous measurement, so it is represented by a point-to-point graph.

WHAT ARE THE KINDS OF SCIENCE?

Historically, science has been divided into two categories: **basic (pure)** and **applied (practical)**. This distinction was based on who was doing the research, why the research was conducted, and how the resulting information was disseminated. Basic research was done by scientists in academic institutions (the public sector), and the resulting information was published in accessible scientific journals. Basic scientists investigated how the natural or physical world works. Applied science was done by scientists working in industry (the private sector) or for the government, and the resulting information was not public knowledge. Applied scientists investigated things that were useful to mankind, such as medicines, cars, detergents, and weaponry. They found new ways to use information produced by basic science researchers . Most of the material goods in production today are the product of the combination of basic and applied sciences.

In many ways, society has benefited from the use of basic research by applied researchers. A basic science researcher discovered that insects attract one another by giving off chemicals (pheromones). An applied science researcher took this information and used it to develop a trap for garden pests, like Japanese beetles. However, all the products developed by applied scientific research have not been beneficial to people. A basic researcher studied how the nervous system of insects works, how messages are sent from the eye to the brain and then to the muscle. This understanding was used by applied researchers to develop nerve gas.

The connection between basic and applied science has become increasingly blurred as new technologies and methodologies are developed. The new areas of biotechnology (using molecular biology to develop new products) and bioengineering (the manipulation of DNA) have, in many instances, merged basic and applied research by having the same researchers work in both areas. Academic researchers, for example, can also work as researchers for chemical or pharmaceutical companies.

With scientists now doing both types of research, questions about conflicts of interest are frequently being raised both by the public and the scientific community. Some questions that you should ask are: Where do researchers' loyalties lie? and Do they have a financial interest in the research? Questions of scientific ethical and moral conduct are matters of growing concern.

Today, scientists debate their responsibility for the ways in which their work is used by society. They, too, have ethical responsibilities as human beings. This means that you will be hearing more from scientists. These issues will affect your life and the lives of future generations. Science today is not a small, limited area of human endeavor; it is part of your everyday existence.

WHAT DO YOU NEED TO KNOW?

Words of warning: Numbers can misrepresent information. As you have seen, the ways in which an experiment is set up, which questions are asked, how the data is analyzed, and how the data are presented can lead researchers to draw different conclusions from the same experiment. Researchers manipulate data when they analyze it. Data can be manipulated so that researchers are shown to their best advantage, or so that it yields the results that researchers expected. One of a scientist's ethical responsibilities is to present information without distortion. No one is completely objective and you must be aware of any biases that might affect how the research was done, how the results were analyzed, and how the conclusions were reached. The only way to know this is by asking questions.

Science, both private and governmental, is big business. It consumes enormous amounts of human effort, money, and time. This may seem unimportant, but think about who pays for this research, either through taxes or by buying the products. The people who decided which research is to be funded and carried out are directly or indirectly accountable to you. Therefore, it is important that you let them know what you want and don't want.

In this text, I will discuss many of today's biospheric problems, how they have arisen, and what can be done to ameliorate their effects on the biosphere, and on you. A biospheric problem is any biotic (living) or abiotic (nonliving) thing that impacts or affects the biosphere in a negative way. Biospheric questions and problems are complex. They have to be responded to now, not left for future generations. You will be asked to make choices in your life about what to do about these problems. To make choices wisely, you must be informed. You must understand what scientists are talking about. You must ask the right questions to find out what is being said and what is *not* being said. You must understand who is saying what and why they are saying it. You must know what information is available and what it really means.

To ask the right questions, you must have a basic understanding of some of the concepts, ideas, and principles that make the **biosphere** (earth) function. You need a scientific base. This book will help you gain that base and find some resources so that you can expand on this base throughout your life.

ASSIGNMENTS FOR PART 1: SCIENCE

1. Explain how two scientists might use the same data and arrive at different conclusions.
2. Explain to your 70-year-old aunt or uncle why it is important to get a second medical opinion before making a decision about a health problem.
3. Explain what *state-of-the-art* means in science. Find a newspaper article that supports your explanation.
4. List two factors that might limit the range of scientific research.
5. Why is it important to remember that scientists are people?
6. Compare basic and applied research. Give an example of each type.
7. Explain to your 10-year-old sister or brother how the scientific method works.
8. Choose something you have seen and propose a teleological and a causal explanation for it.
9. Compare qualitative and qualitative data. Which type of data is preferred by scientists and why?
10. Why do scientists collect many observations when they do experiments?
11. Using either real or imaginary data, draw a graph for a discontinuous characteristic. Be sure to accurately label the X-axis and the Y-axis, and give your graph an informative title.
12. Using either real or imaginary data, draw a graph for a continuous characteristic. Be sure to accurately label the X-axis and the Y-axis, and give your graph an informative title.
13. After your first exam, organize and analyze the class grades. From your analysis, draw at least one conclusion and one inference. Draw a graph to support your conclusion.

SUGGESTED READINGS

Committee on the Conduct of Science. (1989). *On being a scientist.* Washington DC: National Academic Press. (discussion of what it means to be a scientist)

Darwin, Charles. (1957). *Voyage of the beagle.* New York: Dutton. (Darwin's own account of his voyage)

Eiseley, Loren. (1975). The dancer in the ring. In Loren Eiseley, *All the strange hours.* New York: Charles Scribner's Sons. (essay on Darwin as a synthesizer of information and how science works)

Gould, Stephen Jay. (1981). *The mismeasure of man.* New York: W.W. Norton & Co. (a reevaluation of scientific research into IQ)

Hubbard, Ruth. (1990). *The politics of women's biology.* New Brunswick, NJ: Rutgers University Press. (another view of science, feminism in science)

PART 1: LIFE

WHAT IS LIFE?

You know when something is alive, but to actually define *life* is not simple. Sometimes people define life as "not dead," but this definition does not tell us what life *is*, only what it is not. Life is that quality that lets us distinguish living organisms from dead organic organisms or inorganic matter. Living organisms exhibit a set of characteristics. Living organisms can:

- **reproduce:** make more of themselves, either by sexual or asexual means
- **grow and develop:** increase in size and complexity
- **use energy:** take in energy (e.g., sunlight, food) and use it to maintain internal order
- **respond to stimuli:** react to their environment (e.g., light, heat, touch, chemical changes)
- **maintain homeostasis:** regulate and keep a constant internal environment
- **exist in a specific environment:** fit the abiotic and biotic requirements of a specific habitat.

Possessing these characteristics means that something is alive; having only a few of these characteristics means that it is not. Even though a crystal grows and exists in a specific environment, it is not alive. It is inorganic matter.

WHERE DID LIFE COME FROM?

There are several theories that attempt to explain how life began on earth. Some of these theories can be related to science; some cannot. These theories can be grouped into three categories: directed creationism, panspermia, and gradual aggregation.

Theories of **directed creation** say that life was created by a supreme being. There is no way to test the presence of a supreme being directly or indirectly in a laboratory, therefore, these theories cannot be considered to be part of science.

Theories of **panspermia** say that life was created somewhere else, then deposited on earth. Some of these theories assert that life was brought to earth embedded in meteorites. Others assert that life was brought to earth by spaceships. Some of these theories are currently being studied by scientists.

The theory of **gradual aggregation** states that life arose on earth over a long period of time as inorganic material formed into groups (**aggregates**). This is the theory that is most widely accepted by the scientific community today. This theory can demonstrate: how science works, how experimentation can support an hypothesis, how organisms interact with the environment to change it, and how organisms depend on the environment. To help you gain a sense of the geologic time involved in this theory and where *Homo sapiens* enter this chain of events, see Table 1-1.

In 1924, A. I. Oparin and J. B. S. Haldane (1929) proposed the theory of gradual aggregation to explain how life arose on earth. They hypothesized that our solar system is the result of a tremendous explosion (the Big Bang) that occurred about 15 billion years ago. The explosion produced an expanding cloud of dust and large chunks of inorganic matter. Larger chunks of matter attracted smaller chunks of matter. The larger the chunk, the more attracting power it had. The largest chunk in our solar system became the sun; lesser chunks became the planets of our solar system. As the chunks grew in size (mass), they became denser and heated up. The sun ignited about 5,000 million years ago. As the earth coalesced, it became a molten mass. Slowly its surface cooled until precipitation could accumulate on the surface. Bodies of water formed. Earth was about 70° Celsius (158° Fahrenheit) hotter than it is today, so the bodies of water were warm. Oparin and Haldane postulated that the first stable atmosphere was not like the one we have today, but was instead composed of water (H_2O) vapor, ammonia (NH_3), and methane (CH_4), with no free molecular oxygen (O_2).

Table 1-1: A Simplified Time Line of Events

YEARS AGO	EVENT
15 billion	THE BIG BANG
↓	↓
5,000 million	SUN ignites
↓	↓
4,600 million	EARTH coalesces
	ATMOSPHERE accumulates (reducing)
	CHEMICAL EVOLUTION begins
↓	↓
4,000 million	NUTRIENT SOUP (prebiotic)
↓	↓
3,500 million	BIONTS (heterotrophs)
	BIOLOGICAL EVOLUTION begins
↓	↓
2,500 million	BIONTS (autotrophs)
	oxygen produced
↓	↓
1,800 million	OXYGEN ATMOSPHERE finished
	(oxidizing)
	CHEMICAL EVOLUTION ends
↓	↓
65 million	MAMMALS take over
	100,000 Homo sapiens appear
↓	↓
10,000	Last ICE AGE ends
	agriculture starts
↓	↓
5,000	WRITING invented
↓	↓
1,000	Anasazi arrive in the Grand Canyon
↓	↓
500	Europeans migrate to Americas
↓	↓
TODAY	1,000-fold increase in human population

Chemists use a shorthand to represent different elements. Each element is identified by one or two letters, depending on its Latin or Greek name (e.g., hydrogen is H and aluminum is Al). This shorthand is used to show the composition of small and large molecules. For example, a water molecule (H_2O) is composed of two atoms of hydrogen and one atom of oxygen.

Earth's first stable atmosphere was called a **reducing atmosphere** because it contained no free molecular oxygen. The atmospheric gases dissolved in rainwater and accumulated in the warm bodies of water. Minerals and salts from the earth's crust were dissolved in the water and were carried into the standing bodies of warm water. These bodies of water, with their dissolved materials, are referred to as the **nutrient soup**.

The amount of energy in earth's environment, which bombarded the nutrient soup, was greater than it is today. There was no ozone layer to block out most of the ultraviolet radiation (UV rays), there was more electrical activity, and temperatures were warmer. When the energy interacted with the chemicals in the nutrient soup, chemical reactions occurred and, by spontaneous generation, produced small organic compounds (e.g., simple organic sugars and amino acids). These simple organic sugars and amino acids were made not by a living organism but by random physical interactions. These simple organic compounds are said to have been *abiotically synthesized*. The prefix *a* in this word means "not," therefore, **abiotic** means "non-life" or "not living." *Synthesis* means "to bring things together to make something new."

The next part of the theory explains the production of large organic molecules and compounds. Oparin and Haldane hypothesized that if the nutrient soup, with an accumulation of abiotically synthesized simple organic molecules, were to undergo a process of being repeatedly dried, heated, then wetted over a period of time, larger organic molecules called **polymers** would be abiotically synthesized (e.g., proteins). These warm bodies of water would then contain the nutrient soup, simple organic molecules, and large macromolecules, and, because of physical interactions, coacervate droplets would form. **Coacervate droplets** are not living single-celled organisms, but they do have some characteristics that are similar to those found in living single-celled organisms.

Coacervate droplets are a concentration of materials surrounded by a water shell. The water shell acts like a cell membrane, keeping the inside environment separate from the outside and selectively letting only certain things in and out. The droplets increase in size as more material passes through the water shell. This increase resembles growth. When the droplets reach a certain size, they break into smaller droplets. This breaking apart resembles asexual reproduction in living cells.

According to Oparin and Haldane, the coacervate droplets increased in organization and stability until they became **prebionts** (before life). After enough changes had occurred, the first **bionts** (life) were formed. These bionts were single-celled heterotrophs. **Heterotrophic** organisms get their energy from food (organic matter) from the environment. The first heterotrophs absorbed the organic materials they needed from those made abiotically in the nutrient soup.

Until the first bionts were formed, all changes were chemical changes (chemical evolution). The word *evolution* means "change." However, once living bionts were formed, biological changes occurred and biological evolution began. Both chemical and biological evolution continued to occur.

Over time, a new kind of biont arose — one that could take in simple inorganic materials and make its own organic materials. These new bionts were **autotrophs**. *Auto* means "self" and *troph* refers to "feeding or getting energy," therefore, autotrophs are organisms that make their food from inorganic materials. This process produced as a side-product free molecular oxygen (O_2). This oxygen was released into the environment and slowly accumulated to change the atmosphere from reducing to **oxidizing**. Today's atmosphere is composed of approximately 78% nitrogen gas (N_2), 21% oxygen gas (O_2), less than 1% carbon dioxide (CO_2), and other gasses, including carbon monoxide (CO), argon (Ar), helium (He), and neon (Ne). Thus, autotrophs changed their environment.

As the oxygen concentration in the atmosphere rose, some important changes occurred:

- The ozone (O_3) layer was formed, reducing the amount of ultraviolet radiation (UV) reaching earth's surface.
- Chemical evolution ceased because the required materials from the reducing atmosphere were no longer available.

After the oxidizing atmosphere was established, chemical evolution stopped and only biological evolution was possible. Organic molecules could only be produced by living organisms.

What you have just read is a summary of the gradual aggregation theory. You might ask why science accepts this theory. Is there any way to test if it actually explains what happened on earth? Is there any evidence available to support this theory? The answer is yes: Several parts of the theory have been tested, yielding supporting evidence.

Since the 1950s, scientists have been designing experiments to test different parts of the gradual aggregation theory. To do this, research teams placed the hypothesized reducing-atmosphere gases in closed glass apparatus and zapped them with electrical charges. They then analyzed the resulting materials and found that, even after as short a period as a week, simple organic molecules had abiotically formed. These experiments have shown the possibility of abiotically producing simple organic molecules from a reducing atmosphere with sufficient amounts of energy. Experiments have also demonstrated the abiotic synthesis of proteins and coacervate droplets. These experiments point to the possibility that abiotic synthesis and gradual aggregation of organic materials could indeed have happened.

How could you tell the first living single-celled organisms (bionts) from prebionts? The theory describes a continuous process from simple organic molecules to bionts. This process has been broken down into separate parts for discussion. Think of the life cycle of a human being, from birth to death. We break it down into parts to discuss it (e.g., newborn, infant, toddler, child, teen-age, young adult, adult, middle-age, old age). People usually don't look different from one day to the next. We only see change, such as the change from teen-age to adulthood, after a number of small changes have accumulated. We can easily distinguish between an 18 year old and a 35 year old. In the same way, you would not be able to tell the difference between the first biont and the preceding prebiont. Only with an accumulation of changes can we say that one is living and one is not living.

Another part of the hypothesis that has scientific support is the change of the atmosphere from reducing to oxidizing. Remember that the change did not occur overnight, but was a slow accumulation of free molecular oxygen. Bionts that evolved at different times would have been adapted to the environment at that time. Therefore, if an organism arose in a reducing atmosphere, it could only live in a reducing atmosphere; if it evolved in a partially oxidized atmosphere, it would do best with some oxygen present; and if it evolved in an oxidizing atmosphere, it would do best with a high percentage of oxygen.

There are on earth today organisms that can live only in a reducing atmosphere, those that can live in a partially oxidized atmosphere, and those that can only live in an oxidizing atmosphere. These organisms are bacteria.

The existence of bacteria with different atmospheric requirements supports the theory that the atmosphere did change slowly from reducing to oxidizing.

The Oparin and Haldane theory, with a few modifications, is the one that is most widely accepted by the scientific community today because it has the most supporting evidence. This evidence has come from many different disciplines (e.g., geology, chemistry, physics, astronomy, oceanography, biology) and from hundreds of researchers over many years. The testing of this theory is an excellent example of science as a synthetic discipline (using many different kinds of information to support an idea).

HOW DOES LIFE CHANGE?

How life changes is explained in science by the **theory of evolution**. The word *evolution* is used by many people to mean not only that things have changed, but also, incorrectly, that things have improved. For example, a car advertisement may state that a new "breed" of car has evolved. You expect progress in the car's design and performance. However, the term *evolution* means only that there has been *change*, that there are *differences*. Change does not have to mean *improvement*. All other uses of the term misrepresent what is scientifically meant by *evolution*.

The theory of evolution is the binding theory of the biological sciences. It is also an excellent example of science as a synthesis discipline. The name that is most closely associated with the theory of evolution is Charles Darwin, who is credited with proposing and establishing it. However, Darwin was not the only person to propose the theory. Alfred Wallace, a naturalist working in the East Indies, wrote to Darwin outlining his ideas on how organisms change. His ideas were similar to Darwin's. Darwin's and Wallace's ideas were presented to the scientific community in London in 1858. Historically, Darwin has received sole credit for the theory because he had gathered a tremendous body of supporting material, which he published in 1859 (*The Origin of Species*), while Wallace did not have supporting data.

Darwin's theory of evolution explains how species change over time and the mechanism of change (how it happens). Until the introduction of Darwin's theory, the generally accepted explanation for how things changed was the **theory of spontaneous generation**. People thought that, given the correct conditions, organisms spontaneously arose. For example, if meat was left out on a counter for a couple of days, maggots (larvae) could be observed crawling over it. The meat was thought to have spontaneously generated the maggots.

Given the knowledge and technology of the 1800s, the theory of spontaneous generation made sense. There were limits to what could be seen or detected. Tiny eggs, mold spores, or bacteria were not yet detectable.

Darwin read and talked to other scientists in the areas of natural science, geology, and economics, and incorporated their ideas and information with his own observations to synthesize a new idea of how things change. He was not the only naturalist thinking about how species change. In fact, his grandfather had written about this. Darwin was a great synthesizer of information.

Darwin said that species change slowly over time, and, to explain the variation and number of different species in the world in 1859, earth had to be very old. In order for his theory to work, there had to be enough time for the changes to occur. In the 1800s, the church had determined that the earth was only a few thousand years old. This meant that the earth was not old enough for Darwin's theory to work.

However, Darwin had read the works of a geologist named Charles Lyell (1797-1875), who had proposed a theory called **uniformatarianism** to explain how the earth changes, how mountains are built, and how valleys are formed. Lyell's theory stated that the same processes that shape the earth now are the same forces or processes that had shaped the earth in the geologic past. Mountains and valleys were formed by the action of water, winds, and volcanic activity. These activities are very slow, therefore, the earth had to be much older than a few thousand years. Lyell had based his ideas on the work of another geologist, James Hutton (1726-1779), who had stated that earth was molded by gradual processes. Lyell's theory of uniformatarianism gave Darwin the geologic time necessary for his explanation of change to work.

Darwin not only needed time, he also needed to find a mechanism to explain how change occurs. He had read an essay written in 1798 by Reverend Thomas Malthus on human population. In this essay, Malthus suggested that human population size was limited by three forces: disease, war, and famine. Without these control factors, the human population would increase beyond its ability to feed itself. Darwin incorporated Malthus's ideas into his mechanism of natural selection. Thus, scientists depend on the work of other scientists to develop new ideas and technologies.

One reason Darwin's theory gained wide acceptance was that he used many lines of evidence (e.g., fossils, similarities between living species, the domestication of animals). Probably the clearest evidence of change is in the domestication of animals and plants. For example, if you were to start with a flock of sheep in year one and measure the length of their wool, you would have some sheep with long wool, some with short wool, and some with intermediate-length wool. If you could earn more money for long wool, you might want to start a breeding program to increase the number of the sheep in your flock with long wool. How would you do this? You would not allow sheep with short wool to breed. Instead, you would take those males with the longest wool and breed them with the females with the longest wool. The lambs produced would have a variety of wool lengths. The next year, you would repeat the process, always breeding only those sheep with the longest wool. Over a 10-year period, you would produce a flock of sheep with longer wool than you had in your original flock. There would have been a change in the length of wool, a change you could see and document with records. Evolution (change) would have occurred.

Today this same process can be seen in the many breeds of domesticated cats and dogs. It takes only eight generations of selective breeding to produce a new breed of cat that has consistent characteristics that are different enough from the original parents that you would never confuse the two.

Using all the lines of evidence and the ideas of others, Darwin synthesized a new and coherent theory of how species change. Darwin's theory has six basic assumptions: (See Figure 1-3).

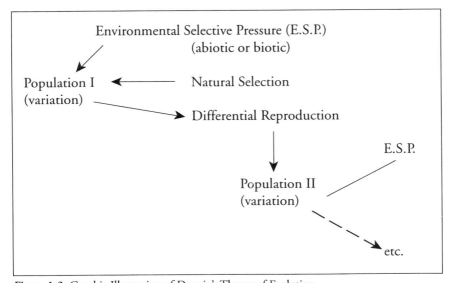

Figure 1-3: Graphic Illustration of Darwin's Theory of Evolution

Darwin's assumptions were that:

1. There will be more individual organisms born in a population than will live to reproduce.
2. There will be **variation** in the population, i.e., no two individuals will be exactly alike.
3. Some environmental factor will influence or affect the population. That factor can be either biotic (living) or abiotic (nonliving) and will act as an **environmental selective pressure** on the population.
4. The process of **natural selection** will occur. The individual organisms in the population that are best adapted to the environmental selective pressure will have a better chance of living to reproduce. They will be naturally selected.
5. **Differential reproduction** will occur. Differential reproduction means that because more individuals that are adapted to the environmental selective pressure are left to reproduce, there will be more individuals born that are adapted to that environmental selective pressure.
6. Populations evolve or change; individual organisms do not. The population becomes adapted to the environmental selective pressure, the individual plant or animal does not. Individuals are, or are not, adapted at birth, they do not evolve to meet the selective pressure.

The following hypothetical example further illustrates the theory of evolution. In a population of 100 rabbits, some have long ears, some short ears; some have long legs, some short legs; some are all white, some spotted; and some have thick coats of fur and some thinner coats of fur. There are many differences between individuals in the population. One rabbit may have long ears, short legs, spots, and a thick coat of fur; another may have long ears, long legs, no spots, and a thin coat of fur. Each rabbit was born by chance with a specific combination, or set, of variations. Each rabbit did not plan which variations it would have at birth.

One year the winter is much colder than normal. The cold temperatures are an environmental abiotic selective pressure that affects the rabbits. Those rabbits that, by chance, are born with thick fur coats will be naturally selected for. They have a better chance to survive the winter and live to reproduce than do the rabbits with the thin fur coats. The beginning population of 100 rabbits had 50 with thick fur coats and 50 with thin fur coats. By the next spring, 60 rabbits have survived the winter: 40 have thick fur coats and 20 have thin fur coats. In the spring, the rabbits mate randomly. Some thick-furred rabbits breed with other thick-furred rabbits, some thick-furred rabbits breed with thin-furred ones, and some thin-furred rabbits breed with other thin-furred rabbits. When the young are born, there will be more offspring with thick fur coats and than with thin fur coats because there were more thick-furred rabbits alive to breed. When more offspring in the next generation of rabbits have thick fur coats and are adapted to the environmental selective pressure of the colder winter, this is called differential reproduction.

The next generation has 100 rabbits, but the number of rabbits with thick fur coats is 70 and with thin fur coats 30. This population of rabbits is better adapted to the cold winter than the previous generation was. If the process were to continue for several years, the number of rabbits in the population with thick fur coats would increase until there were no thin-furred individuals. The population would have changed, or evolved, over time. The individual rabbits had not changed that is, a thin-furred rabbit didn't just decide to grow a thick fur coat because the winter was extra cold. Each rabbit was either adapted to the selective pressure at birth, or it was not.

In this example, the cold winter was an abiotic environmental selective pressure that produced change in the rabbit population. But a biotic environmental selective pressure, such as a new predator (a wolf) moving into the area, could have achieved the same end. What would happen to the rabbit population with a different selective pressure? The variation now being selected for might be for leg length (faster versus slower rabbits) or color (blending with the surroundings versus not blending). Different variations are naturally selected for, depending on the environmental selective pressure. One year it might be the extra cold winter, the next a predator, the next a warmer than normal winter, etc. Evolution is opportunistic it is not goal oriented. Change occurs as a result of the environmental selective pressures that affect the population; these pressures are not predictable. Populations do not change to reach some goal or ideal, they change in response to the environment.

The theory of evolution refers to the "survival of the fittest." This concept has been and still is misinterpreted to mean that the strongest will survive. Historically, it has been used to support and justify many social and political ideas, from colonization of other countries to the western expansion of the United States. The concept of survival of the fittest does not mean that those who are physically strongest will survive, or have the right to survive. The fittest individual organisms are those that are, by chance, best adapted to a specific selective pressure, and therefore, leave the most offspring. Survival of the fittest has nothing to do with which organism is "best," only with which organism leaves the most offspring in the next generation.

The speed at which evolution, or change, occurs depends on the strength of the environmental selective pressure. The stronger the selective pressure, the faster the rate of evolution. An example of the relationship between the strength of selective pressure and the rate evolution is the use of DDT to control body lice. DDT is a man-made chemical that does not occur naturally in the environment. It was used to get rid of body lice during the World War I. It was very effective as an abiotic selective pressure because only those body lice that were, by chance, resistant to DDT could survive to reproduce. Since DDT was a man-made chemical, it was expected that no body lice would survive. However, during World War II body lice were again a problem, and most of the lice were resistant to the level of DDT used the first time.

What had happened? By chance, a few lice were resistant to DDT, and most of the lice that were not resistant were killed, leaving mainly lice that were resistant. The resulting population was resistant to DDT in varying degrees. Thus, DDT was a strong selective pressure and, consequently, there was rapid change in the population. This example illustrates two things of interest:

1. the rate of change is related to the force of the selective pressure and
2. the amount of variation in a population is often unknown.

There is generally a tremendous amount of variation among individuals in a population, and until a selective pressure acts on the population, some of that variation may not be identifiable.

Because genetics and molecular biology were not known in the 1800s, Darwin did not explain how variation arose, he only wrote about observable variation. Today, with our increased knowledge and understanding of how organisms work and how characteristics are inherited, we know that variation stems from mutations, or changes, in the genetic material in all cells. These changes in genetic composition (mutations) are passed on to future generations through reproduction. It is this heritable variation on which natural selection acts.

There are two types of reproduction: sexual and asexual. Sexual reproduction involves the fusion of two cells to produce a new individual organism; each of these two cells contributes to the new organism's genetic composition. In animals, these cells (egg and sperm) are called gametes. In asexual reproduction, only one individual produces the new organism; the single parent consequently contributes all of the genetic information that is transmitted to the new organism. A mutation can be expressed as a new variation in the next generation. If the organism reproduces asexually, the mutation will be passed on to all of the next generation, but if the organism reproduces sexually, the variation, or mutation, may *not* be passed on to the next generation because each parent contributes only half of the offspring's DNA.

The theory of evolution explains how organisms change over time. All scientists agree that organisms change, or evolve. You may have heard the statement that not all scientists believe in evolution. This statement is often made by people who believe in creationism as a way to explain the diversity of organisms on earth. They also claim that creationism is a science. First, it must be made clear that creationism is *not* a science. A science requires that a theory be testable and that repeatable data support the theory. Creationism is a belief. There is no way to test or prove its ideas by experimentation. A belief system is not a science, unless the belief is testable.

The charge that not all scientists believe that evolution occurs comes from the fact that, although all biological scientists accept that evolution occurs, they may differ on the rate at which it occurs. Today in the scientific community, there are at least two major schools of thought on how evolution by natural selection occurs. (See Figure 1-4.)

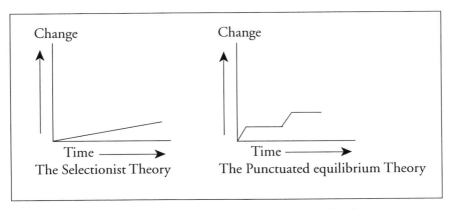

Figure 1-4: The Selectionist and Punctuated Equilibrium Theories of Evolution

One group the **Neo-Darwinists**, or **gradualists** — holds that evolution is a slow, gradual process. The other group holds that evolution is not a slow, gradual process, but instead, that it occurs in two recurring phases — a rapid phase followed by a stable, or equilibrium, phase. This concept is called **punctuated equilibrium**. These two groups agree that evolution occurs by natural selection, but disagree on exactly how it works (the pattern of change). It is thought that some species evolved in one way and that other species evolved in the other way. This doesn't mean that evolution does not occur; it means that organisms can evolve *differently*.

Darwin's theory explains how change occurs, but how do new species arise? The formation of new species is the concept of **speciation**. What is a species? A species is a group of organisms that can breed under natural conditions and produce viable offspring. How do you start with one species and end with two or more species? This part of the theory explains the diversity of organisms in the world today. The formation of two or more species from one species depends on the separation of one population into two or more populations that are unable to continue to interbreed. (See Figure 1-5.)

Once a population is separated into different segments, each segment is subject to different environmental selective pressures. Over time, each new population changes to such an extent that, if individuals from the different populations come into contact with each other, they are no longer able to interbreed. There are now two species where originally there was only one. Historically, in geologic time, the two species had a common ancestor that may or may not have looked like the species that have arisen. The separation can be physical (e.g., the formation of a river, canyon, or mountain range), or biological (e.g., a mutation that makes the individuals no longer reproductively compatible). Speciation has lead to the biodiversity that we see today. Biodiversity refers to the number of different species in an area, an ecosystem, or the earth as a whole.

The concept of speciation has led to many misleading statements about the evolution of *Homo sapiens*. The idea that evolutionary theory states that man descended from the apes is incorrect. What the theory says is that before people, as we understand them, evolved, there was an ancestral organism that may not have resembled man or ape, but from which, over time, two lines of organisms arose. One of these lines eventually evolved into man and another line evolved into modern apes. There is no missing link that looks half-ape half-man.

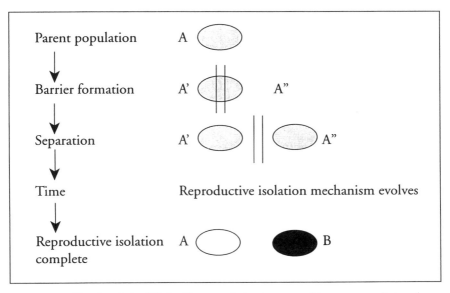

Figure 1-5: How Speciation Works (One population is separated into two populations by a barrier, then evolves separately.)

An understanding of evolution is important because it affects your everyday life. For example, most people have taken antibiotics at some time in their life. The purpose of taking antibiotics is to kill bacteria that make them ill. Antibiotics are a powerful abiotic environmental selective pressure. The bacteria in the system that are not resistant to the antibiotic are killed and the person gets better. But what about the possibility that there may be some bacteria in the population that, by chance, are resistant to the antibiotic? They will be naturally selected and will survive to reproduce. There will be differential reproduction in the next generation of bacteria, and more resistant bacteria will be produced than nonresistant bacteria. This is important because bacteria primarily reproduce asexually. This means that if there is a mutation for resistance to the antibiotic, it will be passed on to the next generation.

What does this mean to you? It means that you may need larger doses of the antibiotic to get rid of a bacterium and, in the future, you may need a new antibiotic because the bacteria are resistant to the one you took at first. Bacteria evolve quickly because they reproduce quickly, by asexual reproduction, and because antibiotics are a very strong environmental selective pressure. Thus, understanding the theory of evolution can help you to make health decisions.

ASSIGNMENTS FOR PART 1: LIFE

1. You are out walking and come across something you have never seen before, how would you decide if it is a living organism?
2. Compare the three categories of theories that explain how life arose on earth.
3. Draw a diagram of the Oparin/Haldane theory of how life arose on earth. This should be usable as a study guide.
4. On your diagram of the Oparin/Haldane theory, indicate where supportive evidence for the theory is available.
5. Compare a coacervate droplet with a single-celled organism.
6. Using the Oparin/Haldane theory, discuss how the biotic component of the biosphere has affected the abiotic environment. Be specific.
7. Discuss how Darwin's theory of evolution is a good example of science as a synthesis discipline. Include at least two specific examples in your discussion.
8. Using an example of your own (real or fictional), explain Darwin's theory of evolution. Be sure to include the vocabulary of the theory in your discussion.
9. Give three examples of possible abiotic environmental selective pressures and three examples of possible biotic environmental selective pressures.
10. What does the phrase "survival of the fittest" mean?
11. Compare the theory of punctuated equilibrium with the neo-Darwinist, or gradualist, theory.
12. What is creationism and why is it not a science?
13. Using an example of your own (real or fictional), explain how speciation occurs.
14. What is the relationship between variation in a population and natural selection?
15. Explain why the statement that an individual organism evolves to meet a selective pressure is false.
16. Explain what is meant by the statement that evolution is opportunistic.
17. Using an example of your own choice, explain how a strong selective pressure affects the rate of evolution.

SUGGESTED READINGS

Eiseley, Loren. (1957). The slit. In Loren Eiseley, *The immense journey.* New York: Vintage Press. (essay on man and geologic time)

Leopold, Aldo. (1966). *A Sand county almanac.* New York: Random House. (naturalist essays on environmental ethics)

Mayr, Ernst. (1976). *Evolution and the diversity of life.* Cambridge, MA: Belknap Press of Harvard University Press. (essays in evolutionary biology)

Wilson, Edward. (1992). *The diversity of life.* New York: W.W. Norton & Co. (issues of biodiversity and the environment)

PART 2: SYSTEMS

HOW DO THEY WORK?

A **system** is a unit of study composed of separate parts that function together. The size of the system to be studied is determined by the researcher and can vary from an individual organism, to a rain drop, to the whole earth. An understanding of the **modern view** of how living organisms are unique is essential to the study of any system. The modern view can be divided into five parts for discussion:

1. All parts of an organism are interconnected, so that if you change one part, you change the whole organism. For example, if you have a vitamin C deficiency, you could develop bleeding gums and loose teeth. These problems, in turn, could lead to other organismic problems.
2. Living systems are constantly changing because of the interactions between internal and external (environmental) forces. For example, if the temperature drops, you may begin to shiver. Your body is responding internally to the external change.
3. Changes in the internal processes of an organism are the result of opposing forces that are constantly shifting emphasis. The forces that build up matter are called **anabolic**; those that break down matter are called **catabolic**. During the lifetime of an organism, both kinds of reactions are constantly occurring, but there is more of one type than another at different times in the life cycle. You can think of these two forces as being on a seesaw. For example, in a young developing organism, the anabolic reactions predominate as new tissues are being formed. During the middle of the life cycle, the anabolic and catabolic process are evenly balanced. At the end of the life cycle, the catabolic process predominate.
4. The whole is greater than the sum of its parts because of the interaction of those parts. This means that the whole organism has characteristics or functions that the individual parts do not have. These new characteristics

or functions are called **emergent properties**. Emergent properties cannot be predicted by studying the component parts of an organism in isolation. These new characteristics or functions are the result of the levels of organization and the interactions of the parts. For example, if you took a car apart, you would have all the parts, but you would not have a functioning car. The car is the result of the interactions of the parts. The car has functions and characteristics that the parts by themselves do not possess. The same is true for living organisms.

5. An accumulation of small quantitative changes in an organism can lead to large qualitative changes. For example, if you heat a kettle of water by raising its temperature one degree at a time (a small quantitative change), the water will eventually boil, then turn to steam. The change of water to steam is a large qualitative change produced by many small quantitative changes. Similarly, a small increase in your blood cholesterol level every year for 10 years could result in the hardening of your coronary arteries, the narrowing of your blood vessels, and high blood pressure. These changes could lead to a heart attack. Thus, a large qualitative change results from an accumulation of small quantitative changes.

The modern view of organisms involves not only the parts that make up an organism and how they interact, but also how the organism changes and what brings about these changes. These same concerns are important when studying any system. When this approach is applied to the study of a system, it is called the **holistic approach**. To understand a system, one must study all of its parts, their interactions, and their relationships.

Science is organized around several main ideas; one of these is an **hierarchical organization**. *Hierarchical* means that there is a set of nested characteristics, or units, with one smaller unit fitted, or nested, into the next larger unit or higher level. A hierarchy is a graded series (e.g., going from smallest to largest or from simplest to most complex). When you study a system, you need to know at which level of the hierarchy you are working. Where does the system fit? The levels of organization are:

- subatomic particle
- atom
- molecule
- organelle
- cell

- tissue
- organ
- organ system
- organism
- population
- community
- ecosystem
- biome
- biosphere

As you progress from a lower level to a higher level, complexity increases and new properties arise. These new properties or characteristics (emergent properties) are produced from the interaction of the units. Different laws work at different levels of organization. All levels are interconnected and interdependent. A cell, for example, has properties that separate organelles lack; a forest has properties that a tree lacks. The cell and the forest have characteristics produced by the interactions of their parts. When you know at which level of organization you are working, you can relate the system to its place in the organizational hierarchy.

Systems are not isolated entities. Since an hierarchy is really a continuum, any system studied must have boundaries. These are artificially set by the scientist. This means that if you want to study a pond ecosystem, you can set the boundaries at the edge of the pond. But even though you can assign a boundary at any point, you must remember that the pond is only part of a larger system, and that the larger system will impact on the pond. Thus, knowing what is going on in the watershed will help you understand how the pond functions.

You can also think of yourself as a system. Imagine that you are ill and must see a doctor. The doctor may do routine tests and discover that you have high blood pressure. To treat your problem, he or she must understand not just your body, but also your life. There will be questions about home life, job, leisure activities, friends, etc. You are not an isolated organism. To treat you, the doctor must understand where and how you fit into your world. Your high blood pressure may be caused by some organic problem; it may also be caused by a stressful job. Finding out the cause of your high blood pressure will help your doctor treat your problem, not just your symptoms. To study any level of the hierarchy, you must ask the right questions. If you ask the wrong questions, you may not get useful information.

Systems are complex and maintain their stability or equilibrium by using **homeostatic mechanisms**. These mechanisms are **feedback mechanisms**. There are two types of feedback: positive and negative. If a feedback mechanism is positive, the product of a reaction keeps the reaction going. If the feedback mechanism is negative, the product of the reaction slows down or turns the reaction off. Your household thermostat is a negative feedback mechanism. When the temperature reaches the desired level, the furnace is turned off; when the temperature falls below the desired level, the furnace is turned back on. The temperature is the product and it determines whether the process continues.

What would happen if the thermostat worked as a positive feedback mechanism? The furnace would stay on because the rise in temperature in the room would tell the furnace to keep working. Most systems in nature maintain their homeostasis by using negative feedback mechanisms. (Feedback mechanisms will be discussed again in Part 2: Ecosystem.)

In this book, you will study the larger hierarchical levels beginning with the organism and continuing up to the biosphere. To understand the ideas in this book, you need a basic vocabulary:

- *organism:* an individual living unit. An organism belongs to a **species**. A species is a group of individuals who are capable of interbreeding successfully under natural conditions. Successful breeding means that the offspring are viable and are able to reproduce at maturity. For example, a donkey and a horse can interbreed. The resulting offspring is a mule. Mules are almost 100% sterile (not able to reproduce), therefore, donkeys and horses are two different species.
- *population:* a group of individuals of the same species living in the same area
- *community:* all the organisms of different species living in a designated area ecosystem: all the organisms in an area, and their environment, functioning as a unit
- *biome:* a large regional unit composed of ecosystems that have similar structures and functions. It has a predominate type of vegetation that is adapted to a particular environment (e.g., a grassland or tropical forest).
- *biosphere:* all of the earth's ecosystems and communities on a global scale

We, as human beings, are animals. Thus, we are part of the hierarchy. We are part of the system; what we do affects everything else and everything else affects us. It is important not to lose sight of where you fit into this hierarchy. It is also important not to forget how intricate and wonderful all systems are.

ASSIGNMENTS FOR PART 2: SYSTEM

1. Explain to your 10-year-old (fifth grade) neighbor the modern view of how living systems work. Be sure to use examples that he or she will understand.
2. Explain how the holistic approach to studying systems relates to the modern view.
3. Using a real or an imaginary example, explain hierarchical organization.
4. Using real or imaginary examples, explain positive and negative feedback mechanisms.
5. Why are African and Asian elephants considered to be two different species?

SUGGESTED READINGS

Starr, Cecie, & Taggart, R. (1995). Biology: The unity and diversity of life (7th ed.). New York: Wadsworth Publishing Company. (biology textbook)

Stein, Sara. (1993). Noah's garden: Restoring the ecology of our own backyards. New York: Houghton Mifflin Co. (how the backyard garden works as an ecosystem)

PART 2: BIOSPHERE

WHAT IS IT?

The biosphere is all of earth's ecosystems and communities on a global scale. To understand how it works, you must understand both its buffer strip and its abiotic parts and their interactions. The biosphere, surrounded by the atmosphere, is composed of the hydrosphere and the lithosphere. (See Figure 2-1.)

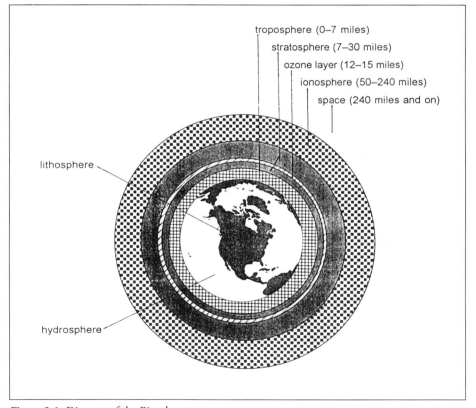

Figure 2-1: Diagram of the Biosphere

Earth's **atmosphere**, which is composed of gases and particulate matter, extends from the surface of the earth up about 240 miles. It is divided into several layers, which vary in composition and temperature. The **troposphere** is the lower level of the atmosphere; it extends approximately seven miles up from the earth's surface. The troposphere contains 99% of the water vapor and 90% of the air in the atmosphere. This is the part of our atmosphere that is responsible for all weather. The **stratosphere** is the atmospheric level above the troposphere; it extends up 30 miles above the earth. The **ozone layer** is important in protecting the earth from the sun's high energy ultraviolet (UV) radiation; it is located in the lower level of the stratosphere. The outer layer of the atmosphere is the **ionosphere**, which is composed of ions (charged atoms). The ionosphere extends from 50 to 240 miles from the surface of the earth. Space begins at the outer limit of the atmosphere.

The biosphere is composed of the **hydrosphere**, which refers to all the water on the earth, and the **lithosphere**, which refers to all of the soil and rocks. Approximately 71 percent of the earth's surface is covered with water and 29 percent is land. In discussing the biosphere, scientists often use terms such as acre, hectare, kilometer, and meter. To appreciate fully what is being discussed, you should understand the following terms. (See Table 2-1.)

Table 2-1: Terms of Metric Measurement and Their U.S. Equivalents

Metric Term		U.S. Equivalent
kilometer (km)	=	6/10 of a mile
meter (m)	=	1 yard or 3 feet
acre	=	43,560 square feet or the size of a football field
hectare (ha)	=	2-1/2 acres or 2-1/2 football fields

See Appendix II for a more complete list of equivalent measurements.

Where does life exist on earth? Although some lichens and bacteria are found on the top of Mount Everest, which is six miles high, most organisms are found only about one mile into the atmosphere. Some organisms live below the earth's surface: the deepest place in which life has been found (so far) is in the Marianas Trench, which is 7.5 miles deep. Thus, life on earth is contained in a layer 8.5 miles thick. In that layer there are now approximately 36 million living species. The exact number of species is constantly changing because new ones are discovered and others become extinct. The number of species alive on earth today is about 1% of all the species that have lived on earth since life arose approximately 3.5-million-years ago.

The biosphere, which is made up of all of these ecosystems and communities, is often compared to a spaceship. Like a spaceship, the biosphere is self-contained and has its own life-support systems. Everything that the crew on a spaceship will need must be aboard before lift-off because nothing new will be available once the mission begins. The biosphere, too, is self-contained and has its own life-support systems.

It is a super ecosystem with many feedback loops and interactions, all of which are needed to maintain homeostasis. It is a highly integrated, self-organized, and controlled system. The biosphere is **bioregenerative** in that the plants, animals, and microorganisms generate materials, recycle materials, and control life necessities. They are responsible for maintaining the biosphere's life-support systems. These systems are complex and interactive, and we do not fully understand how they function or how they are maintained. The support system is a vast, diffuse network of processes that work on a geologic time scale that is different from the human time scale of 80-100 years.

One hypothesis proposed to describe the biosphere is the **Gaia hypothesis**, named after the Greek goddess "Mother Earth." Proposed by James Lovelock, this hypothesis points out that the biosphere is self-regulating and that the biotic part of the biosphere plays an important role. It actively interacts to modify and control the chemical and physical conditions of the biosphere.

Because we do not fully understand the workings of the biosphere's systems, we cannot correct them if they malfunction. Only in recent history has man come to understand how complex the biosphere is and how little we actually know about it. It has become apparent that we have the capacity to damage the biosphere and also that we may not be able to correct the problems that occur. This realization has led to the **precautionary principle**, which states that before any plan or new technology is used, it must be tested to prove that it does not harm the biosphere. In other words, safeguards must be built into proposed plans or any new technology before it can be used. The necessity for this principle can be seen in the damage done by DDT, the problems that surround the use of radioactive nuclear material, and the problems that result when wetlands are converted into housing developments.

The materials produced by the life-support systems of the biosphere are essential to life on earth. These materials can be classified as **market** or **nonmarket** goods and services. Market goods and services are those that are sold for money (e.g., food). Nonmarket goods and services are provided free by the natural environment (e.g., clean air, clean water, good soil). Nonmarket goods and services are necessary for the production of market goods and services. Historically, nonmarket goods and services were taken for granted. This practice led to the degradation of the environment and to damage to these goods and services. Today, although many nonmarket goods and services are not sold directly to you, you pay for them indirectly, such as when businesses have to clean up the environment. Electricity costs more when power plants must remove pollutants from the smoke they emit before it can be released into the atmosphere.

The terrestrial ecosystems that cover 29% of the earth's surface are important to a functional biosphere. These systems can be divided into three environmental types: **fabricated (developed)**, **natural**, and **cultivated (domesticated)**. Fabricated ecosystems are man-made. They are powered by fossil fuel and are energy intensive. They produce large amounts of waste, pollution, and heat. Cities and highway corridors are examples of fabricated systems. Highway corridors are man-made environments that develop along highways (e.g., strip malls, commuter housing). Natural ecosystems are systems that have not been altered by human activity. Natural systems are self-maintained and are based on solar power. Cultivated systems have been modified by human activity. They are managed to increase production for human benefit. Agricultural systems are examples of cultivated systems. Cultivated systems are solar powered and energy subsidized. They depend on energy subsidies such as fertilizers and pesticides to increase their production over natural systems. Cultivated systems produce pollution problems because of their extra energy subsidies. (See Table 2-2.)

Table 2-2: Terrestrial Systems in the United States

Environmental Type	Area of U.S. Occupied	Energy Usage	Ratio
Fabricated	6%	34%	5.6
Cultivated	24%	27%	1.12
Natural	70%	39%	0.56

The pollution produced by fabricated systems causes extensive damage to cultivated and natural systems. Cultivated systems produce problems for natural systems, too, but of the three environmental types, fabricated systems have the most impact on the biosphere. The biosphere's life support systems are maintained by natural and cultivated systems.

HOW DO WE STUDY IT?

Ecology is the discipline that has been developed to study the biosphere. The term ecology comes from the Greek words *oikos*, meaning "house or place to live," and *logos* meaning to "study." Literally, ecology means the study of organisms at home. The field of ecology is the study of organisms (populations or communities), their interrelationships, and their interactions with the environment. Scientists look not only at the structure of an ecosystem, but also at its function.

Ecology is one of the many biological sciences. **Biology** studies life in all of its manifestations. If you think of biology as a layer cake, there are two approaches you could use to divide it. (See Figure 2-2.)

You could divide the cake into layers or slice it into pieces. The layers (basic divisions) represent things that are common to all life (e.g., anatomy, morphology, genetics, physiology). The slices (taxonomic divisions) represent specific kinds of organisms (taxonomic divisions) (e.g., botany, zoology, bacteriology, ornithology). Ecology is a basic division because all organisms are adapted to their environment and to their part in an ecosystem. In terms of hierarchical organization, ecology is concerned with those levels above the organism.

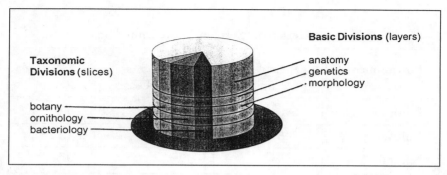

Figure 2-2: Various Approaches to Studying Biology

Ecologists (scientists who perform ecological studies) study ecosystems to understand how they are structured and how they function. When discussing the biosphere, there are two concepts that are important: **value** and **objectification**. In today's culture, the word value most often refers to monetary value: how much is it worth or how much could it be worth. This is very often how ecosystems or parts of ecosystems are discussed. You have probably heard the tropical rainforest valued in terms of how much the plants might be worth to future pharmaceutical advances, or of how much money can be made if the forests are cut for lumber. There are other meanings to the word value, too, and we need to talk about what other "values" ecosystems might have besides how much can be made from the lumber, the plants, or the gold in the ground.

People also talk about the *aesthetic* value of natural ecosystems in today's stressful society. But again, aesthetic value can be translated into tourist dollars.

Do ecosystems have other "values"? Are these "values" equally important, or even more important, than the money they can generate? Yes, ecosystems do have other "values" — they maintain the biosphere. Old-growth forests provide a habitat for organisms, purify air, purify water, act as a sponge to prevent flooding, regulate local climate, and stop soil erosion. Tropical rainforests are crucial in maintaining productive tropical soils and the global climate by removing greenhouse gases. When making decisions about natural ecosystems, one must weigh all of the "values," not just the monetary values.

Since we cannot reconstruct ecosystems, we must think carefully before we destroy them.

One reason mankind has caused so much disruption in the biosphere is that we no longer see ourselves as part of nature, as animals, as heterotrophs, as consumers, as part of a food web, and as organisms that affect our ecosystem. We do not see how our actions affect the workings of the biosphere and how that in turn affects us. One way that we have distanced ourselves from our natural place in the biosphere is through objectification. **Objectification** is the process of turning things into objects, often as consumer goods. We eat hamburgers, not cows; we buy redwood decking lumber, not 1,000-year-old trees. By turning things into objects, we create an emotional distance that allows us to separate ourselves from responsibility. We do not acknowledge that these objects were living, functioning parts of their ecosystems. We do not consider what impact the removal of the organisms may have on the ecosystem and the biosphere. Objectification allows us to act irresponsibly.

In 1933, Aldo Leopold published an essay titled "The Land Ethic," in which he argued that mankind has an ethical responsibility to the land (i.e., the biosphere). Leopold is seen as the founder of the **land ethic** concept. He argued that people have an ethical responsibility to maintain and improve the natural environment. They do not have the right to treat the biosphere as an object to be bought and sold. Man, as part of the biosphere, has a responsibility to not abuse and mistreat the biosphere. The lack of an ethical, integral approach to the biosphere and man's separation from nature are responsible for the damage that man has done to the biosphere.

ASSIGNMENTS FOR PART 2: BIOSPHERE

1. How large is your bedroom in feet, yards, and meters?
2. How many football fields does a parking lot cover if is covers three acres?
3. Explain to your friend how the biosphere is like a spaceship. Be sure to include an understanding of the term *bioregenerative*.
4. Compare the three environmental types of terrestrial systems based on energy type, energy usage, and impact on the biosphere.
5. Construct a graph that represents the information given in Table 2-2. Be sure to give your graph an informative title.
6. Explain how the discipline of ecology relates to biology.
7. Explain to a friend how the land ethic relates to the concepts of value and objectification.

SUGGESTED READINGS

Krebs, Charles. (1994). *Ecology* (4th ed.). New York: Harper-Collins. (ecology textbook)

Starr, Cecie, & Taggart, R. (1995). *Biology: The unity and diversity of life* (7th ed.). New York: Wadsworth Publishing Company. (biology textbook)

PART 2: ECOSYSTEM

WHAT IS IT?

The field of ecology is the study of organisms (populations or communities), their interrelationships, and their interactions with the environment. Scientists look not only at the structure of an ecosystem, but also at its function. The term *ecosystem* was first used in 1935 by Arthur Tansley to describe an organized unit. It includes the biotic and abiotic parts of a unit and their interactions.

BIOTIC COMPONENT

General Categories of Organisms

An ecosystem has two major biotic components: producers and consumers. **Producers** (e.g., green plants) are autotrophs that are self-nourishing. They convert radiant energy into usable chemical bond energy by the process of photosynthesis. Producers take in inorganic substances, such as water and carbon dioxide, and use radiant energy to convert them into carbohydrates.

Consumers (e.g., animals, bacteria, fungi, microorganisms) are heterotrophs who obtain their energy by eating other organisms. Consumers are divided into four groups depending on their energy (food) source. This classification system is based on function (what the organism does to get energy), not on species.

- **Herbivores** (e.g., rabbits, cows, sheep), sometimes referred to as grazers, obtain their energy by eating producers.
- **Carnivores** (e.g., vultures, spiders, wolves), sometimes referred to as predators, obtain their energy by eating other animals, such as herbivores or other carnivores.

- **Omnivores** (e.g., bears, raccoons) consume both producers and other heterotrophs.
- **Saprovores** (e.g., fungi, bacteria) obtain their energy by consuming decaying organic matter.

The movement of energy from the producers through the saprovores — in other words, who eats whom — is called a **food chain**. Energy moves from grass to rabbit to fox. Most organisms eat more than one thing, and so are part of several food chains. The linking of the biotic parts of ecosystems into a network of energy transfers is called a **food web**. (See Figure 2-3.)

HOW DO WE FIGURE OUT HOW IT WORKS?

Ecosystems are open systems that materials enter and leave. They are not self-continued, and they are very complex. To study these complex systems, ecologists often use models. To build a model to mimic a system, ecologists gather information about both the biotic and the abiotic parts of the systems, including how they interact and interrelate. Models help scientists to gain a better understanding of the system and to make predictions about the system. They are useful because they can be manipulated without permanent change to the natural ecosystem.

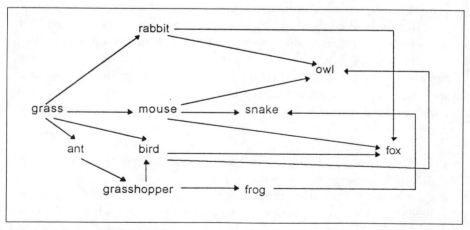

Figure 2-3: A Simplified Food Web

Using a model, scientists can alter part of a system (e.g., add, delete, increase, or decrease an element) and see what might happen as a result. For example, if you constructed a model of a river and its surrounding watershed, you could predict what would happen to the river and the watershed if you were to build a dam on the river. To fulfill its function, a model must be as complete as possible. The more information used to build the model, the better will be the predictions it allows. There is a saying that a model of an ecosystem is like going to a restaurant and eating the menu. A model gives you a good idea of how the ecosystem is structured and functions, but it is not the same as the real ecosystem. Eventually you have to work with the natural system.

There are many different ways to construct models, but one method uses the energy language devised by H. T. Odum. This method uses seven symbols in constructing a model. (See Table 2-3.)

Table 2-3: Odum's Energy Language Symbols and Their Meanings

SYMBOL	MEANING
○→	**energy source** represents the source of energy for the system and the direction of flow
⊥	**heat sinks** represent the loss of usable energy to the system as each organism carries out cellular respiration
⊐	**bullets** represent the producers
⬡	**hexagons** represent the consumers
◇	**storage bins** represent stored materials, e.g., water, oxygen, and nitrogen
⟩⟩	**arrows** represent interactions between organisms
—→	**lines** represent links between components of the system and show the direction of movement, e.g., between the water storage bin and the producers

After a model is constructed, each part must be quantified, that is, number values must be determined. The scientist would need to know how much energy is passed from the producers of the systems to the consumers and how many of each different kind of organism are present. This information is collected by ecologists when they research the natural ecosystem (**field work**). Since not everything in an ecosystem can be measured — you cannot count every bacterium in the soil — sampling is done. Models are constructed based on the information obtained in samples. Remember that models are only as good as the information used to construct them, therefore, the more information you have the better your model will be. (See Figure 2-4.)

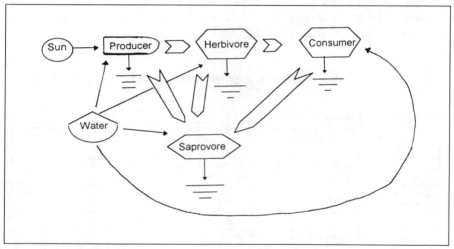

Figure 2-4: A Simplified Ecosystem Model Using Odum's Energy Language

Population Dynamics

Population dynamics is the study of how the size of a population changes over time. The formula used to determine change is: $I = (b - d) N$. I is the rate of change in the number of individuals in the population, b is the average birth rate, d is the average death rate, and N is the number of individuals in the population at the present time. If the birth rate rises above the death rate, the population will increase; if the death rate rises, the population will decrease.

Any natural factor or cultural factor that affects the birth rate or the death rate will affect the change in population size (e.g., improvements in sanitation facilities decrease the death rate from disease).

If you graph the growth of a population over time, the population will have one of two types of growth curve: J-shaped or S-shaped, although these are extremes and most populations show a growth curve somewhere between the two. Growth curves are named after the shape of the graphed data. The **J-shaped** (see Figure 2-5) curve resembles the letter J.

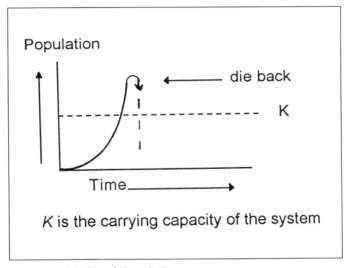

Figure 2-5: A J-Shaped Growth Curve

Populations that exhibit J-shaped growth curves are **density-independent**. The size of the population does not limit the growth rate of the population. J-shaped growth-curve populations operate with a positive feedback mechanism: the more individuals there are in the population, the more are born and the population grows at an increasing rate. The population increases exponentially; surpasses the carrying capacity (K) of the environment; increases in size until it comes up against an inhibiting, or limiting, factor, such as running out of territory or food (**environmental resistance**); then dies back or crashes. The **carrying capacity** (K) is the number of individuals (maximum density) that a system can support without damaging the resources of the system.

Very few populations exhibit J-shaped growth curves. An example of this kind of population is the lemming, a small mouse-like rodent that lives in the arctic. The old saying "like lemmings to the sea" is derived from their cyclic mass migrations out of their normal territory. After a migration, the population falls drastically. Why do lemmings migrate?

Lemmings live in a cold environment with a short growing season where little plant life is produced annually. The lemmings obtain the nutrients they need by eating the plants, but as the lemmings' population increases, more and more of their environment's nutrients are locked up in the lemmings. Over a period of years, the nutritional quality of the plants decreases and the lemmings, even though they are eating enough food, are getting less and less nutrition. When the nutrition level falls far enough, physiological changes occur in the lemmings that trigger a massive migration. The population crashes.

Fresh water algae also exhibit the J-shaped growth curve. At some time during the growing season (spring through summer), you may notice that a lake turns bright green for a short while (an algal bloom), then returns to its normal color. The color change results from a rapid increase in the population size because of excess nutrients in the water. When the algae run out of nutrients, the population crashes or dies off. The die back for a J-shaped growth curve can wipe out an entire population.

The **S-shaped curve** (see Figure 2-6) resembles the letter S, or has a sigmoid shape.

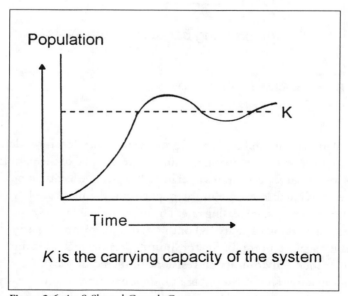

Figure 2-6: An S-Shaped Growth Curve

Populations that exhibit a S-shaped growth curve are **density-dependent**. As the population density increases, the rate of growth decreases. The size of the population limits the growth of the population. The larger the size, the smaller the growth rate. The population increases until it reaches the carrying capacity of the ecosystem, then stabilizes around the carrying capacity.

The carrying capacity (K) is not fixed in a system; it varies over time because of changes in the biotic and abiotic parts of the system. For example, an increase in winter precipitation (abiotic factor) could result in an increase in a producer population (biotic factor) in the spring, which would allow the rodent (seed eater) population to increase. Thus, the K for an organism may be high at one time and lower at another.

There are two levels of carrying capacity: maximum (subsistence) density and optimum (safe) density. **Maximum (subsistence) density**, is the maximum number of individuals that can eke out a living in a system without damaging the system's resources. **Optimum (safe) density** is the number of individuals that can live in a system without damaging it, but at a lower level of density than the maximum. This is a more secure level of density. If an ecosystem's carrying capacity were to decrease, the population at the optimum density would be affected less than a population at the maximum density. Historically, carrying capacity has been expressed simply as a number of individuals, without considering if different individuals have different impacts on the system. Today, the concept of carrying capacity includes the maximum number of individuals and the intensity of use that can be supported by a system without degrading the environment or damaging its resources.

Populations increase in size as individuals reproduce. Species are identified as having one of two strategies for reproduction: K-selected or r-selected. K-selected (equilibrium) strategists exhibit S-shaped growth curves. These species live in ecosystems with a constant, predictable climate, have long life-spans, mature slowly, and reproduce continuously for several years after maturity. Individuals are generally large and the population has a slow growth rate (e.g., mammals, birds, oaks). They exhibit **quality** reproduction: they produce a few offspring at a time and expend a large amount of energy raising them to reproductive age (e.g., whales, elephants, humans).

Opportunistic (r-selected) strategists exhibit J-shaped growth curves. These species live in ecosystems with unstable and variable climes, have short life-spans, mature early, and reproduce once. Individuals are generally small and the population has a rapid growth rate (e.g., insects). They exhibit **quantity**

reproduction: they produce large numbers of offspring and expend little or no energy raising them to reproductive maturity (e.g., frogs, locusts, algae).

Human beings are naturally a K-selected species, but due to cultural changes, they have switched to a r-selected strategy. (See Figure 2-7.) This change in strategies has had a tremendous affect on the biosphere. The human population, which had reached 5.4 billion in 1992, is increasing at a rate of almost 2% per year. Many people believe that all the current biospheric problems are directly or indirectly caused by human overpopulation. (See Figure 2-8.)

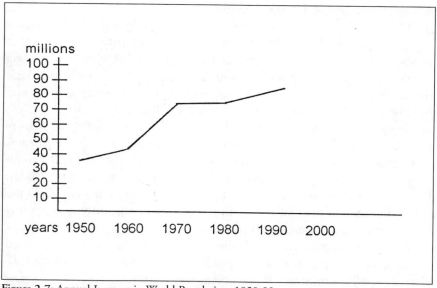

Figure 2-7: Annual Increase in World Population, 1950-93

The human population has already surpassed the carrying capacity of the biosphere. Remember that K is concerned not only with the maximum number of individuals, but also with the intensity of use of the resources by those individuals. It is not only our numbers that surpass the carrying capacity of the biosphere, but also how we consume resources and the technologies we use to acquire those resources. The effects of this overpopulation will be discussed throughout the remainder of this book.

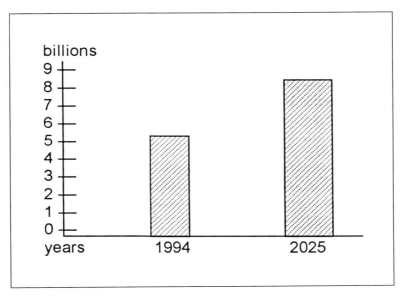

Figure 2-8: World Population Size, 1994-2025

The Effects of Overpopulation

The increase in human population can be attributed to a decrease in infant mortality, a decrease in the death rate for all age groups, and an increase in length of life. These changes are the result of new knowledge in medicine, public health, and education.

As the human population increases, more food is required. This, in turn, affects the whole biosphere. For example, we have increasingly turned to the oceans and rivers for sources of protein. The size of the ocean led many to believe that its resources were inexhaustible. New fishing technologies have enabled us to increase the amount of fish harvested, and, since the late 1950s, the amount of fish caught has increased fourfold. This has resulted in the exploitation (**overfishing** and **overharvesting**) of these resources. The amount of fish caught has begun to steadily decrease, and for some species there is the very real possibility of extinction. Of the 65 species of fish assessed by the National Marine Fisheries Services of the United States, 18 species are overfished. The Atlantic tuna breeding population has fallen from 250,000 to 22,000. The decreases in fish catch are the result of the combined factors of pollution, habitat destruction, and overfishing.

An excellent example of overfishing is the salmon. By the summer of 1994, the salmon population was so depleted that the threat of total collapse was imminent. In an effort to rebuild the salmon stock, a ban was placed on salmon fishing in the Pacific Northwest. One can find the same result by looking at the demise of the oyster business in the Chesapeake Bay, (See Figure 2-9.), the loss of fishing in the New Jersey estuary system, and the near-extinction of whales.

The problem is how to use the resource (fish) to help feed the human population without damaging the fish populations. This is the concept of **sustainable yield**. Overpopulation places increased demands on all of the resources of the biosphere. The management of world-wide resources requires international cooperation.

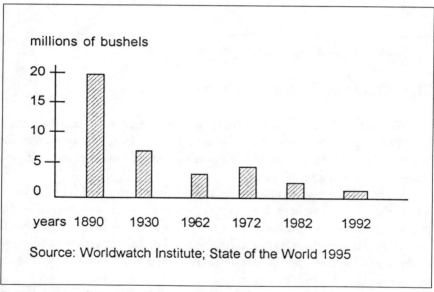

Figure 2-9: Annual Oyster Catch on the Chesapeake Bay

Population size or density is naturally controlled by disease, war (predation), and famine. This concept can be illustrated by examining the relationship between a predator and its prey. To illustrate the predator-prey relationship, look at how rabbit and fox populations interact to control population size. (See Figure 2-10.)

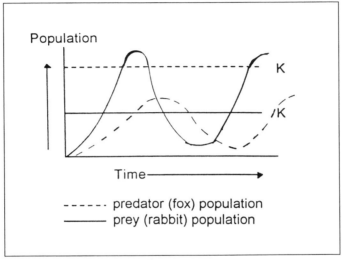

Figure 2-10: Predator-Prey Cycle

Both species are *K*-selected strategists and exhibit S-shaped population growth curves. The number of foxes is related to their food supply (the rabbits), and the number of rabbits is related to the number of predators (the foxes). Each species fluctuates around the carrying capacity for that species. If the rabbit population increases because there is more grass to eat due to good weather, more offspring will live to maturity. As a result, the fox population will increase because there is more food, and more fox pups will also live to maturity. An increased fox population will consume more rabbits, and the rabbit population will decrease. With fewer rabbits to eat, the fox population will then decrease. This means that more rabbits will live to reproduce, and the cycle will start over. The increase in the fox population lags behind the rabbit population increase. These interactions keep both populations stable around the carrying capacity for that system.

What would happen if the predator population greatly decreased or if the predator was removed? The removal of the population-control mechanism would allow the prey population to increase greatly until it ran into some other environmental resistance factor. For the rabbit population, the environmental resistance factor might be the lack of plant material to eat. Overpopulation would result from the removal of the control agent.

Predators act as biotic selective pressures on the prey species. They eliminate the old, the diseased, and the less well-adapted. Predators keep the prey population in good health. When the predator population falls to very low levels or is removed from the area, the prey population suffers. For example, man has removed or eliminated most of the top predators from the Eastern deciduous forests in the United States. As a result, the deer population has exploded. The increase in the deer population has caused numerous problems. It has placed a tremendous stress on the producers that the deer eat. There are not enough producers and the deer population is in poor health because of poor nutrition. With the predators gone, the diseased animals are not removed, so they spread disease to more individuals. In addition, in their search for food, deer have come into closer contact with the human population as they move into yards to eat shrubbery.

In the forest, the deer have created a **browse line**. They have eaten everything from the ground up to as high as they can reach. Animals that eat food closer to the forest floor are affected by the increased deer population, too. Remember that all parts of the ecosystem are interrelated and that everything is affected by a change in the ecosystem's predator status. Natural control systems work well because they have evolved over a long period. Predator-prey relationships are finely tuned so that neither species is wiped out.

The Effects of Population Control Efforts

Some species are considered by humans to be pests (e.g., mosquitoes bite us, insects eat the food we grow, beetles devour our flowers). Insects consume or spoil one-third of the food produced annually. Man has tried to eliminate them by using chemicals and machines. We have forgotten that each pest is an integral part of a food chain, a food web, and an ecosystem. When we eliminate a pest, we disrupt the whole system. Let's look at some of our efforts to control pests.

We have tried to eliminate pests by using machines (bugwhackers), insecticides and pesticides (chemical poisons), and traps. Bugwhackers are machines that attract and kill flying insects. Bugwhackers are not selective, so they eliminate not only mosquitoes, but other insects as well. How does using a bugwhacker affect the ecosystem of your yard? Insects are a major food source for many birds. By killing them, you may reduce the amount of food available to the birds. Less food results in fewer birds. Fewer birds can result in an increase in other populations (e.g., tent caterpillars).

Another way to control pest populations is through the use of man-made chemicals (e.g., DDT malathion). These chemicals are sprayed onto plants to kill pests. However, the chemicals do not stay on the plants, but wash off into the soil and into the water systems. Although more specialized chemicals are now being produced, many other organisms are eliminated or stressed when chemicals are used in the environment. Soil organisms are necessary to keep the soil fertile and are at risk from these chemicals. The use of chemicals to control pest populations has decreased the fertility of the soil and its productivity. Thus, the effort to increase the amount of food produced by controlling pest populations has actually decreased the amount of food available.

Other methods of pest control are not destructive to the ecosystem. The use of natural predators to control pest populations is called **biological control** (e.g., ladybugs, lacewings, praying mantises). These are naturally occurring or imported species that, if given the right environment, will live in the system and reproduce the next year. The use of natural chemicals to attract pests to traps is another way to control pest populations. These naturally occurring chemicals are called **pheromones**. These pheromones are chemicals that insects emit to attract the opposite sex. Putting pheromones in a trap attracts specific types of insects to the trap, where they are eliminated. Pheromones are specific, so that only the targeted pest population is affected. The Japanese beetle traps that are set out in gardens use pheromones. **Sterilization** is another way to eliminate a specific pest. When sterilized males are released into the ecosystem, they mate with fertile females and no offspring are produced. Sterilization programs have been used with great success. For example, the larvae of the screwworm fly were injuring and killing livestock, but the sterilization program eliminated the problem.

Another method of controlling pest populations is the practice of **companion planting**. Some plants naturally give off chemicals that repel pests, while other plants attract insects so that they will not eat the food species. By growing these companion plants around a food crop, the pests are either attracted or driven away from the food crop.

Genetic engineering is also useful as a method of pest control. New varieties of plants are being developed that are more resistant to disease and pests.

Growing concern and understanding of the effects of man-made chemicals on the biosphere has led to the concept of **integrated pest management (IPM)**. IPM uses several methods to control pest populations. Chemicals that are specific to the targeted pest are used sparingly and only at specific times in the life cycle of the pest. The use of biological controls and other methods greatly decrease the need for man-made chemicals. Organic farmers and gardeners do use not man-made chemicals.

The Effects of Introduced Species

We have seen that the removal of a natural predator may result in an increase in the prey population, but what would happen if a species were to be placed in an ecosystem in which it had no natural predators? As humans move around the world, they take other species with them, either intentionally or unintentionally. These species are not native to the new ecosystem; they are **introduced** species. They have no natural predators in the new environment and prey on organisms that have not developed defense mechanisms against them. Introduced species can have catastrophic effects on their new environment.

The Hawaiian Islands are isolated, and the plants and animals that have evolved there are unique. There were no rats or pigs on these islands until they were introduced by man. Because these species have no natural predators, their populations have grown unchecked. Pigs destroy the natural forest ecosystem by foraging for food and rats prey on small animals and birds. Since Hawaiian species evolved without these predators (selective pressures), they are not adapted to them and have no defenses against them. Many native species have become extinct because of these two introduced species.

Native Hawaiian birds, too, have been affected by introduced species. Nonnative species have been imported and have brought with them avian malaria. The native birds have no defenses against this disease, and it is driving many species to extinction.

Rabbits were introduced into Australia as a food source, but when some escaped the species became a problem. Again they had no predators and their populations grew and decimated the native plants.

But animals are not the only species that can cause problems if they are introduced into new environments. Introduced plants also can cause problems.

The kudzu vine in the southeastern United States has covered whole forests; loosestrife has taken over wetlands in the Northeast, driving out the native plants; and the water hyacinth has clogged the waterways of the Southeast. Many plants can reproduce asexually, which allows their populations to increase more rapidly than those that depend entirely on sexual reproduction. Foreign fungi and blights have destroyed many tree species. The chestnut and Dutch elm are gone and the red pine is being attacked. These introduced species cause disruption of the natural ecosystem. Once an introduced species has gotten out of hand, it is often impossible to correct the damage it causes to the ecosystem.

One reason introduced species do so well is that they out-compete native species for resources (e.g., land, breeding territories, food sources). Competition can lead to the removal of one of the two competing species. **Gause's principle** states that two organisms cannot occupy the exact same space or niche. This is also called the **principle of competitive exclusion**. The more overlap between the needs of two organisms, the more competition there will be. As the human population increases, it increasingly comes into direct competition with other species for living space or for needed resources. Because of its numbers and technology, the human species can out-compete every other species on the planet.

Where an organism lives in the ecosystem is called its **habitat**. The habitat of a frog is a pond, a downy woodpecker's habitat is the Eastern deciduous forest, a desert tortoise's habitat is the desert, a penguin's habitat is the Antarctic.

The role of an organism in the environment is called its **niche**. The concept of niches covers all aspects of the organism's life and its interactions with the environment. The niche includes where it lives (its habitat), what it eats, how it behaves, and how it functions as part of the ecosystem — EVERYTHING. Understanding the niche of an organism is difficult. Many current biospheric problems are the result of not fully understanding the niche of a species, and therefore not being able to predict what effect removing that organism or altering the environment may have on the system as a whole. For example, if you plant a willow tree in your yard, the water table will fall and the area become drier. As the soil dries out, some species will no longer be able to exist. The milkweed species requires moist soil and will disappear. The monarch butterfly lays its eggs on the milkweed plant so that its caterpillars will have that specific food source as they develop. No milkweed, no caterpillars, and therefore, no monarch butterflies. The willow changed the habitat. Habitat loss, as you can see, has very serious implications. It not only changes the way an area appears. It can also lead to the extinction of a species.

Thus, it is very important to know how all of the pieces of an ecosystem function together. Ecologists hope that by studying all of these pieces and learning how they interact with one another, we can avoid making alterations to earth's ecosystem that will harm our planet, and all of its inhabitants — us.

ASSIGNMENTS FOR PART 2: ECOSYSTEM
BIOTIC COMPONENT

1. Using Odum's energy language, construct a model of an ecosystem you are familiar with, e.g., the Eastern deciduous forest, the Arizona desert. Choose a specific organism to represent the producer and each of the four consumer groups. Include storage bins for oxygen, carbon dioxide, nitrogen, phosphorus, potassium, and water.
2. Explain to a friend why two models of the same ecosystem may give different predictions.
3. Construct separate food chains for what you ate for dinner.
4. Construct a food web for everything you ate today.
5. What effect would a national vaccination program have on the population dynamics of the nation? Why?
6. Discuss the similarities and differences between the J-shaped and the S-shaped growth curves.
7. Explain carrying capacity (K) to a friend.
8. Why is the understanding of K important in today's world?
9. Compare K-selected and r-selected strategies.
10. Using an example of your choice (real or imaginary), explain how predator-prey interactions control population size.
11. What is IPM and how does it work?
12. Explain Gause's principle and how it relates to K.
13. Explain to a friend the concepts of habitat and niche.
14. Using an example of your choice, explain why and how an introduced species affects an ecosystem.
15. Using an example of your choice, explain the statement "human overpopulation is the direct or indirect cause of current biospheric problems."

SUGGESTED READINGS

Krebs, Charles. (1994). *Ecology* (4th ed.). New York: Harper-Collins. (ecology textbook)

Miller, G.T., Jr. (1994). *Living in the environment* (8th ed.). Belmont, CA: Wadsworth Publishing Co. (environmental science textbook with case studies)

Odum, Eugene. (1971). *Fundamentals of ecology* (3rd ed.). New York: W.B. Saunders, Harcourt Brace & Co. (ecology textbook)

Worldwatch Institute. *The state of the world.* New York: W.W. Norton & Co. (annual update of current world environmental issues and what needs to be done)

ABIOTIC COMPONENT

The abiotic component of the ecosystem is made up of everything in the biosphere that is non-living (e.g., wind, temperature, solar input, amount of nitrogen in the soil). It determines where and how living organisms exist. A change in the non-living (abiotic) part of the environment directly affects the biotic (living) part of the system.

Limiting Factors

Why do you find different species of trees living in the forests of Canada, New Jersey, and North Carolina? Why do you find elephants in Africa and not in Norway? Why do species live where they live and not all over the earth? Organisms live where they can get what they need to survive: energy, nutrients, water, sunlight, and the proper temperature range.

Species adapt to a specific environment by the process of evolution. Abiotic physical and chemical factors act as environmental selective pressures. Organisms live where they are adapted to the environment's physical, chemical, and biological factors. Where they live is limited by what they can tolerate. This is referred to as their *tolerance range.*

Organisms have certain tolerance limits for physical and chemical factors. They can survive only within a restricted range for each factor. If the amount of the physical or chemical factor is too low, the organism will be absent from that ecosystem. If a minimum amount of the factor is present, there will be a few of the organisms in the system, but they will be infrequent and will be living under physiological stress. If the amount of the physical factor is well within the range of tolerance for the organism, they will live there in abundance and in good physiological condition. If there is too much of the factor, again, the organism will be infrequent and under physiological stress. If there is an overabundance of the factor, the organisms will be absent because they will be unable to tolerate the high concentration. (See Figure 2-11.)

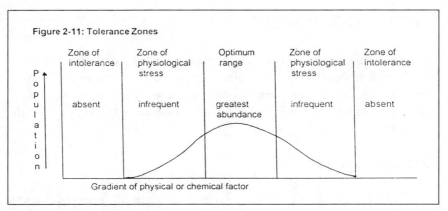

Figure 2-11: Tolerance Zones

The **Limiting Factors Principle**, or the **Law of Tolerance**, states that too much or too little of one or more factors will eliminate an organism from an area. For example, for more than 40 years the lakes in Saskatchewan, Canada, were stocked with small-mouth bass for sport fishing. The hatchery-raised bass were brought into the lakes, grew to maturity, and lived from year to year — but a breeding population was never established. Even though the bass spawned each year, more hatchery fish had to be added to replace those caught the previous year. Why? The temperature of the lake was 60° F, and the adult fish could live and spawn, but a temperature of 65° was required for the hatchlings to develop. The lake temperature was too low, or below the tolerance limit, for the fish. This example points out how the law of tolerance works and demonstrates that different limits may exist for the same factor at different times of the life cycle of the organism. The adult bass had a wider range of temperature tolerance than did the hatchlings. To understand the tolerance limits of an organism you must study its whole life cycle, not one stage.

For autotrophs, the major chemical and physical limiting factors are nutrient availability, the amount of sunlight, the temperature range, and the amount of water available. It is easy to see the effect that changes in limiting factors have on species composition. Look at the vegetation changes as you hike up a mountain. You'll notice that the temperature and moisture gradients change: it gets cooler and drier. The tree species composition changes as you hike through the physical factor gradient. You will see trees that need warmer temperatures and more moisture in the lower elevations (e.g., deciduous tree species). You will see those that require colder temperatures and can tolerate less moisture at the higher elevations (e.g., evergreen tree species).

The major limiting physical factors for both terrestrial and aquatic ecosystems are light and temperature. The major chemical limiting factor for terrestrial systems is water; for aquatic systems it is the amount of dissolved oxygen. Other important limiting chemical factors for both terrestrial and aquatic systems are the amount of available nitrogen and phosphorous.

Changing abiotic environmental factors can cause the removal of an organism from that system because it cannot tolerate the changes that occur. For example, many lakes in Canada, the northeastern United States, and Europe have become more acidic over the past 50 years. The change in acidity is shown on the pH scale, which measures the amount of hydrogen ions in the solution. The pH scale extends from 0 to 14, with 7 as neutral. Below 7, the amount of hydrogen ions increases and the system becomes more acidic. Above 7, there are fewer hydrogen ions and the system is more basic, or alkaline. Most organisms have a narrow range of tolerance of pH: from about 6.8 to about 7.2.

The increase in lake acidity is the result of **acid rain**, which is caused by the nitrous oxides and sulfur oxides in the atmosphere. These materials are produced by industrial processes, which form what is commonly called **air pollution** when they are emitted into the air. When it rains, these pollutants are washed out of the atmosphere and into aquatic ecosystems, where they form acids and lower the systems' pH. As the acidity rises above the tolerance level of the phytoplankton, they die. This starts a reaction throughout the food chain. Species that feed on phytoplankton starve, the lake food chain collapses, and the lake dies. Many lakes are now dead because they are so acidic that nothing can live in them.

Acid rain also increases the acidity of soils. The change in soil pH can stress the soil organisms that are essential to the nutrient recycling processes that maintain good soils. Changing the concentration of a physical factor can have far-reaching effects on the entire ecosystem.

Organisms can be categorized as either **specialists** or **generalists**, depending on their ranges of tolerance. Specialists have narrow ranges of tolerance (e.g., orchids, ferns). Generalists have wide ranges of tolerance (e.g., ragweed, poison ivy, dandelions). What you commonly call weeds are generalists; they can tolerate wide gradients in physical and chemical factors. Many species are sensitive to changes in the environment and quickly reflect these changes.

These species (**ecological indicators**) are useful in indicating if the environment is changing. They help ecologists assess the health of an ecosystem.

Some indicator species are generalists and some are specialists. Lichens are widely used to study air pollution because they are specialists with a narrow range of tolerance for air pollution. As the amount of air pollution increases, the number and health of lichens decreases. Ragweed has been used as an indicator in the West for overgrazing and to gauge the amount of disturbance in national forests. Ragweed is a generalist indicator species with a wide range of tolerance. The more disturbed an area is, the more ragweed there will be.

Similarly, amphibians (frogs, toads, and salamanders) are disappearing rapidly worldwide due to changes in their environment. Because amphibians have both an aquatic and a terrestrial stage in their life cycle, they experience changes in more than one habitat. Amphibians are used as indicators of chemical changes caused by pesticides, herbicides, heavy metals, ozone decline, acid rain, temperature change, loss of wetlands, and the introduction of species. All of these changes lead to decreased amphibian populations.

Energy

The strategy of living species is to persist. To do this, each species must do three things:

1. adapt to the environment by evolution,
2. secure the necessary materials (nutrients) to synthesize the matter they need to construct and maintain the physical structure of life, and
3. secure the necessary energy to run the basic life processes. Without a continual input of energy, systems stop functioning.

All systems are continually becoming disorganized or disordered. The disorder of a system is called **entropy**. One characteristic of living organisms is that they can create and maintain high internal order. Organisms translate their environments into themselves when they take in energy and materials and make biological molecules. Living organisms must trap energy and materials to drive back disintegration reactions. To demonstrate entropy, let's consider a car. To a keep a car functioning properly, you must continually put energy into it, in the form of oil, gas, parts, labor, paint, etc. Without this input of energy, the car would cease to function and would slowly turn into a pile of rust.

Living organisms also must constantly fight against increased entropy. They do this by using energy. You must continually replace skin cells, for example, as the old ones die and slough off.

Energy is the ability to do work within a system. Energy is measured in units called **calories**. A calorie is the amount of heat necessary to raise 1 gram or 1 milliliter of water 1 degree Celsius (C) starting at 15° C.

The study of how energy moves though a system (**energy flow**) is the cornerstone of ecological investigation. Because energy connects all the parts of an ecosystem, studying energy flow can show you how the parts of a system are interconnected. Energy is also a universal feature of ecosystems, therefore, it can be used to describe, compare, and classify ecosystems.

Today's understanding of energy behavior is based on the two **laws of thermodynamics**. Since understanding how energy flows within and through a system is essential to understanding the structure and function of an ecosystem, you will need to understand the laws of thermodynamics. The first law of thermodynamics, the **law of conservation**, says that energy can be transformed from one type to another type, but that during the transformation no energy is destroyed and no new energy is created (e.g., radiant light energy can be changed into chemical bond energy, but the total amount of energy stays constant). Energy is conserved during energy transformations. This is true for all types of energy except atomic energy, in which more energy is produced because mass (material) is converted into energy.

According to this law, you could convert 100 units of light energy into 100 units of chemical bond energy. But this doesn't happen. Why? The second law of thermodynamics (**law of entropy**) states that no transformation of energy is 100% efficient, and that some energy will be changed into a dispersed form of energy (e.g., heat energy). Dispersed energy is not usable to a system. For example, if you started with 100 units of radiant energy and transformed them to chemical bond energy, you would produce only 10 units of chemical bond energy and 90 units of heat energy.

Heterotrophs take in high energy (food) that has stored chemical bond energy, use that energy to carry on all the necessary processes, and give off low energy (heat). The process by which the chemical bond energy is made available to do work for the organism is called **cellular respiration**. All organism autotrophs (plants) and heterotrophs (animals) use cellular respiration to make chemical bond energy available to do work in the organism and give off heat energy. All life is accompanied by energy changes and depends on a continuous flow of energy.

ᵌergy enters the earth as radiant energy from the sun and is captured by ᵌophs through the process of photosynthesis. Radiant energy is converted into chemical bond energy and stored in the organic matter of autotrophs. All the rest of the biosphere obtains its energy from the organic substances made by autotrophs through photosynthesis. All life depends directly or indirectly on the continuous flow of radiant energy from the sun and its conversion by autotrophs.

Ecology is concerned with how light energy is related to ecosystems and how it is transferred and transformed within the system. The behavior of energy in an ecosystem is called **energy flow** because energy travels one-way through the system. It enters as radiant energy and leaves as heat energy. The sun is a thermonuclear furnace that is constantly producing vast amounts of energy in various forms (**radiation**). (See Figure 2-12.)

Radiation is also called **electromagnetic energy**. Electromagnetic energy travels as waves. The length of the wave (the distance between the crests of the waves) increases from gamma rays to radio waves. Different wavelengths of radiation have different amounts of energy; the amount of energy decreases with increased wave length. The longer the wavelength, the less energy it contains. For example, UV rays are short-wave length and have high energy,

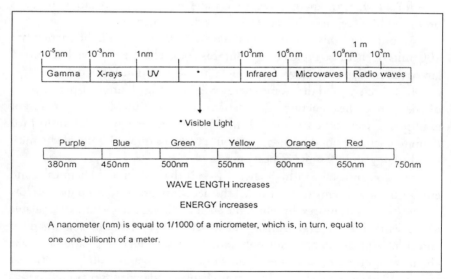

Figure 2-12: The Electromagnetic Spectrum

while radio waves are long and have little energy. Of the complete range of radiation produced by the sun, we see only a portion, called visible light. Some species (insects) can see energy in the ultraviolet range (UV) and some species (cats and pit vipers) can use part of the infrared ranges of the spectrum.

Not all of the solar radiation produced toward the earth actually reaches earth's surface. About 35% is blocked by the atmosphere, and there is further reduction by clouds, water vapor, pollutants, and vegetation. Very little of the incoming solar radiation is used for photosynthesis. Radiant energy is essential for the biosphere in that it is necessary for the process of photosynthesis to occur. (See Table 2-4.)

Table 2-4: Dispersion of the Biosphere's Annual Solar Radiation

Reflected	30%
Directly converted to heat	46%
Used in evaporation, precipitation	23%
Used in winds, waves, and currents	0.2%
Used in photosynthesis	0.8%

Energy interacts with matter in one of three ways: absorbed (taken in), reflected (bounces off), transmitted (passes through). (See Figure 2-13.)

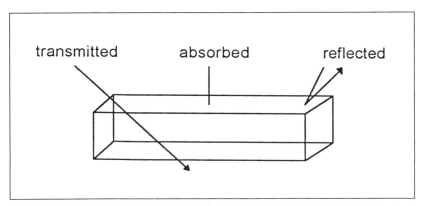

Figure 2-13: Interactions of Radiation with Matter

The absorption of energy in the visible light spectrum results in the colors we see. A substance that absorbs light energy is a **pigment**. The color of an object is determined by the wave lengths that it absorbs (e.g., a red balloon absorbs all the wavelengths in the visible spectrum except red). (See Figure 2-14.)

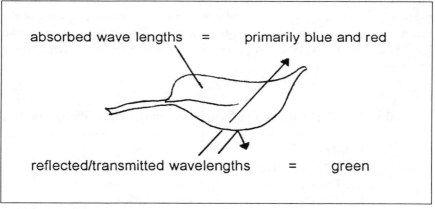

absorbed wave lengths   =   primarily blue and red

reflected/transmitted wavelengths   =   green

Figure 2-14: Green Leaf

If an object absorbs all the wavelengths of visible light it will appear black; if it absorbs none it will appear white.

When energy is absorbed, matter heats up and the heat is radiated into the environment. Any surface or object above 0° C experiences temperature changes (**thermal flux**) (e.g., soil, water, clouds, vegetation, you). (See Table 2-5.) Thermal flux is a **condition of existence** for all organisms. It is an environmental selective pressure that must be adapted to (e.g., animals that live in the desert have evolved behaviors that are adapted to the high day temperatures and the cold night temperatures). Organisms are adapted to the thermal flux of their environmental; therefore this is a limiting factor. For example, when the temperature of the outflow of water from a nuclear power plant is slightly higher than the normal water temperature of the river, large fish kills can result. The change in water temperature, or the thermal flux, may be too high for the fish to tolerate.

Table 2-5: Ecosystems that Experience the Most and the Least Daily Thermal Flux

MOST	LEAST
desert	tropical rainforest
alpine	deep water

There are two important chemical pathways that deal with energy and make it available to living organisms, **photosynthesis** and **cellular respiration**. (See Table 2-6 for a summary of these two pathways.)

Table 2-6: Summary of Photosynthesis and Cellular Respiration

PHOTOSYNTHESIS
$CO_2 + H_2O \rightarrow C_6H_{12}O_6 + H_2O + O_2$
Green plants start with low energy, inorganic molecules—carbon dioxide (CO_2)—and water. A pigment (**chlorophyll**) in the leaves absorbs the blue and red wavelengths and, through a complex series of steps, produces a high energy, organic carbohydrate ($C_6H_{12}O_6$), water, and oxygen. Radiant energy is stored in the chemical bonds of the carbohydrate.

CELLULAR RESPIRATION
$C_6H_{12}O_6 + O_2 \rightarrow CO_2 + H_2O + HEAT + ENERGY$
Cellular respiration is carried out by plants and animals. During the process, high energy, organic, carbohydrate ($C_6H_{12}O_6$) is broken down through a series of steps to release the stored chemical bond energy. Some of the released energy is made available to the organism to do work. The end products of the process are a low energy, inorganic molecule (CO_2), water, and heat.

The solar radiation that is received by green plants (autotrophs) is important to the structure and function of the biosphere. The wavelengths of visible light that plants use to run photosynthesis are blue and red. Anything that decreases the amount of blue and red wavelengths that reach the green leaves affects the amount of photosynthesis that is carried out by the plant.

The energy available to do work in an ecosystem is discussed in terms of productivity. **Productivity** is the amount of organic matter fixed by the process of photosynthesis in a given area over a period of time. In other words, how much carbohydrate was fixed per day by the autotrophs in a 10-acre forest. Productivity is broken down into two categories: **primary productivity** and **net community productivity**. Primary productivity is carried out by the autotrophs of the system and is divided into **gross primary productivity** and **net primary productivity**. **Gross primary productivity** is the total amount of carbohydrate fixed in the systems per area per day; **net primary productivity** is the amount of carbohydrate fixed that is stored in excess of autotroph respiration costs. The net primary productivity is the energy that will be available to the heterotrophs in the system. For example, if you ran a grocery store, your gross profit would be all the money you took in for the week. It would not tell you how well the businesses was doing. To find that out you would have to calculate your net profit. To find this amount, you would have to add up all your costs (e.g., electricity, employee wages, health benefits, water, taxes) and subtract them from your gross profit.

The same is true for autotrophs. They have energy expenses (cellular respiration) that must be subtracted from their gross primary productivity to determine how much energy is stored in the chemical bonds of their organic compounds. This stored energy can be passed on to the rest of the food chain as plant biomass (material) (e.g., fruits, flowers, pollen, wood, leaves, roots). Some autotrophs are more productive than others.

Net community productivity (**secondary productivity**) is the amount of energy stored at the consumer level after the cost of heterotrophic cellular respiration has been subtracted. Ecosystems can be compared by examining how productive they are. Estuaries, mangrove swamps, and tropical rainforests are some of the most productive ecosystems, while deserts and alpine systems are some of the least productive.

All autotrophs carry out the process of photosynthesis, but they all don't do it the same way. Different autotrophs evolved under different environmental stresses, or selective pressures, (different amounts of light, water, and temperatures), so they evolved different photosynthetic processes. Plants grow where their needs for light, temperature, water, and nutrients are met. The two kinds of photosynthesis evolved by higher plants (angiosperms and gymnosperms) are called C_3 and C_4. The names reflect the number of carbons that are found in the first stable compounds in the chemical pathways. (See Table 2-7 for a summary of these two kinds of photosynthesis.)

Table 2-7: Summary of C_3 and C_4 Photosynthesis

	C_3	C_4
peak production	moderate light moderate temperature adequate water	high light high temperature water efficient
inhibited	high light high temperature inadequate water	
examples	wheat, rice, potato	corn, sorghum, sugar cane

The different kinds of photosynthesis are important because they have important food implications. C_4 plants are two to three times more productive than C_3 plants. Most of the world's cultivated primary productivity (food) comes from C_3 plants. With the human populations rising and food productivity not keeping pace, scientists are constantly trying to find ways to make plants more productive. They are trying to breed plants for human use or consumption that can store more energy.

High rates of production occur in natural and cultivated ecosystems when the physical factors are favorable, when the plant's needs are met. Productivity refers to the richness of the ecosystem. A **rich** ecosystem stores energy at a rapid rate. A rich system may have larger quantities of organisms present than a less-rich system, but not always. Richness cannot be determined by counting or weighing the number of organisms present at one point in time. To determine the richness of an ecosystem, you must consider the relationship between its **standing crop** (what is actually there if you weighed it), harvesting procedures, and productivity. For example, if you were to look at a pasture being grazed by cattle, you would have a small standing crop of producers (grass), a high harvest rate, and a fast turn-over-rate: high community productivity.

Primary productivity is not uniformly distributed around the world; some areas are more productive than others. Only one-fourth of the land area of the world has nutrient-rich soil, adequate water, and the right temperatures. Therefore, only one-fourth of earth's land is suitable for high primary productivity, or high food production. Those countries that cannot produce enough food to sustain their population must import food. The acreage needed to grow imported food is call **ghost acreage**. High food production is possible in Europe, North America, and Japan. Thirty percent of the world's population lives in these developed countries, where high food production is possible; 70% lives in countries where food production is only one-third to one-fourth of the amount produced by the top producing countries.

Crop production can be increased if auxiliary energy is pumped into the system. This extra energy can be supplied in many forms (e.g., increased nutrients [fertilizers], increased water [irrigation], use of machinery, lack of insect competition [insecticides], and lack of plant competition (herbicides). These added energies are called **energy subsidies**. The United States uses 10 times as much energy per acre as underdeveloped countries and, as a result, has doubled or tripled its production. Crop yield can be doubled by using 10 times as much energy. There is not a one-to-one correspondence between the amount of energy used and the amount of crop increase per acre.

In the 1970s, scientists using genetic selection produced new varieties of crops that had a higher ratio of edible parts to nonedible parts. The new high-yield plants led to what was called the **green revolution**. These high-yield varieties needed auxiliary energy in the form of fertilizer, water, and pest control to produce the high yields, but world food production increased. Many people thought that high-yield crops, with their energy subsidies, were the answer to feeding the increasing world population.

The United States produces more than 40% of the world's harvest of corn for feed and food. From 1940 to 1985, there was rapid growth in corn yield because of the new high-yield or high-intensity varieties and the use of energy subsidies, but since 1985 there has been little to no rise in harvest. (See Figure 2-15.)

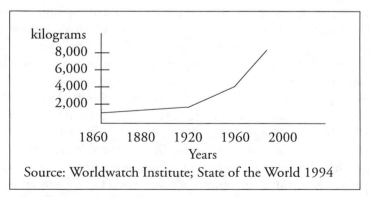

Figure 2-15: United States Corn Yields per Hectare, 1866-1993

Food production has barely kept pace with population demands, and most of the increase over the past few years has been due to increasing the amount of land in production, not to increasing the amount of yield per acre. The process of photosynthesis has built in biological limits and further increases in energy subsidies will not increase productivity. The last year to see large increases in productivity with increased fertilizer applications was 1984. (See Table 2-8.)

Table 2-8: World Grain Production and Fertilizer Use, 1950-1993 (production and usage in millions of tons)

Year	Grain Production	Increment	Fertilizer Use	Increment
1950	631		14	
1984	1,649	1,018	126	112
1989	1,685	36	146	20
1993	1,719	34	130	-16

Source: Worldwatch Institute: State of the World 1994.

Similar trends are seen in wheat and rice yields worldwide. With the growing world population and grain yield not increasing to match, the amount of grain per person is rapidly falling. (See Figure 2-16.)

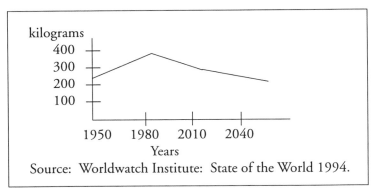

Figure 2-16: World Grain Output Per Person, 1950-1993, with Projections to 2030

Applying more fertilizer has little or no effect above the physiological limits of the plant. Increases above that amount do not increase production. Because food yield will not continue to increase and the population level will, there must be some changes. The changes will have to come in dietary patterns. (See Table 2-9.) It takes two kilograms of grains to produce one kilogram of poultry, four for pork, and seven for feedlot beef. (See Figure 2-17.) In 1994, of the 1.76 billion tons of grain consumed worldwide, 665 million tons (approximately 38%) were used as feed for livestock.

Table 2-9: Per Capita Grain Use and Consumption of Livestock Products in Selected Countries, 1990

Country	Grain Use	Beef	Pork	Poultry	Mutton	Milk	Cheese	Eggs
United States	800	42	28	44	1	271	12	16
Italy	400	16	20	19	1	182	12	12
China	300	1	21	3	1	4	--	7
India	200	--	0.4	0.4	0.2	31	--	13

Source: Worldwatch Institute: State of the World, 1994

Other solutions are being sought to increase food production, including increasing the amount of land under food production, and using aquaculture and hydroponics. Since the most productive areas are already under production, people have turned to the deserts and the tropical rainforests to produce food. To put desert areas under agricultural production requires the input of energy subsidies in terms of water. Although this has been done in many deserts areas (e.g., Israel, the United States), the long-term results have created serious problems. (See section on the water cycle.)

As population demands have increased in areas containing tropical rainforests, these have been harvested for lumber, and then used for agriculture. However, converting tropical rainforests to agricultural land is not feasible because the soils are nutrient poor, erode easily, and are compacted by heavy machinery. These soils only produce food for three or four years before

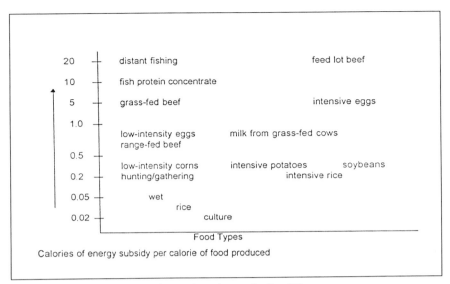

Figure 2-17: Energy Subsidies for Food Production by Food Type

becoming unusable. Thus, destroying tropical rainforests to produce food not only doesn't work in the long term, it also causes other problems (e.g., a loss of biodiversity and an increase in the greenhouse gas carbon dioxide). (See the section on the carbon cycle.) However, food can be produced in a mature tropical rainforest by planting perennial crops within the forest. (See the section on recycling.)

Although it does produce food, **aquaculture**, the farming of fish, requires energy subsidies in the form of grain for fish food. The organisms are vulnerable to climate change, disease, and water pollution. Fish farms are often established in coastal areas that are essential for the maintenance of the ocean fish populations. Mangrove swamps and marsh areas act as nurseries for many saltwater fish, and when the nurseries are destroyed, the ocean populations decrease.

Hydroponics is the raising of plants in a nutrient water solution. These "farms" are greenhouses with plants suspended in containers of water that contain all the necessary nutrients for the plant to grow to maturity and produce a food crop. Hydroponics requires large energy subsidies in the form of climate control and fertilizers. All of the above food production methodologies do produce food, but they cannot produce enough food to feed the growing world populations and all of them generate other environmental problems.

The Food Chain

The transfer of energy through a system from plants (autotrophs) through a series of organisms is called a **food chain**. Remember that because of the second law of thermodynamics, these energy transfers are not 100% efficient. Therefore, food chains are usually limited to four or five links, or steps. The shorter the food chain, the more energy is available per unit eaten or consumed. The closer an organism is to the producer level, the greater the amount of usable energy. (See Figure 2-18.)

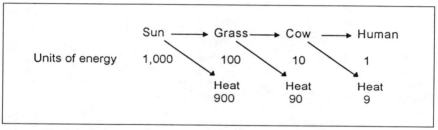

Figure 2-18: Food Chain Energy Transfers and Heat Loss

Organisms can be classified in terms of where they get their energy in the food chain. This is a **trophic classification** (energy). All the organisms that get their energy at the same step in the food chain are on the same **trophic level** (energy level). (See Table 2-10.)

Table 2-10: Food Chain Trophic Classification

Level	Name	Energy Source
First	Producer	autotrophs use radiant energy
Second	Primary Consumer	herbivores are heterotrophs that eat autotrophs
Third	Secondary Consumer	primary carnivores eat herbivores
Fourth	Tertiary Consumer	secondary carnivores eat primary consumers

The last carnivore in the food chain is called the **top carnivore**. Energy flows from one trophic level to the next. An organism that eats both autotrophs and heterotrophs is an **omnivore**. A single organism may occupy more than one trophic level depending on its diet (e.g., man is a primary consumer when eating vegetables and a secondary consumer when eating meat). Because, as a rule, nothing eats humans as a regular diet, man is a top carnivore.

There are two kinds of food chains: **grazing** and **detritus**. Autotrophs are the beginning of the grazing food chain. The **grazing food chain** goes from autotrophs to herbivores to carnivores, for example, grass → cow → man grass → rabbit → fox.

The **detritus food chain** begins with dead organic matter, which is consumed by microbes (organisms that are visible only with a microscope). The primary consumers are bacteria and fungi. These microorganisms are consumed by detritovores (detritus feeders), and these are, in turn, consumed by secondary consumers. Because the organisms involved are small and the detritus food chain occurs primarily underground, it is relatively invisible to the ordinary observer.

In a Northwestern forest, the grazing pathways have approximately 143 species of birds, reptiles, and mammals, while the detritus pathway may have as many as 3,500 microscopic soil dwellers, such as mites, microscopic spiders, millipedes, centipedes, tiny beetles, etc., and a total of 8,000 species recycling the forest litter. In a teaspoon of good topsoil, you will find 1 million single-celled organisms (amoebas, paramecium), 20 million fungi, and 5 billion bacteria. Though you are usually unaware of the detritus food chain, it is crucial to the functioning of the biosphere. Decomposers get their energy from all levels of the trophic structure. It is the detritus pathway that is primarily responsible for the cycling of essential nutrients. (See the section on nutrient recycling.) Of the two kinds of food chains, or pathways for energy to move through the environment, the detritus pathway moves up to 90% of the energy.

Food chains are interlocking because most organisms are part of several food chains. These interlocking food chains form patterns called **food webs**. The intricacies of food webs are difficult to grasp. The following examples may help make this clearer. Removing any one species from an ecosystem can have disastrous effects on the whole system. A simple marine arctic food chain is: phytoplankton → zooplankton → cod → ringed seal → polar bear.

The polar bear, as part of many food chains, is an integral part of the arctic wood web. It is the most abundant large mammal in the arctic, an omnivore, and a top carnivore. Its diet includes the ribbon seal, bearded seal, harp seal, hooded seal, beluga whale, narwhal, musk ox, walrus, hare, goose, birds' eggs, seaweed, tundra berries, and carrion. Other top carnivores (e.g., the fox, raven, gull), depend on the polar bear's leftovers. The complexity of the food webs means that the removal of any one species can easily affect other species in the ecosystem.

When looking at a single organism, it is difficult to know how important it is to other ecosystems, both as a source of food and as habitat. The oak species is used as a food source by 96 species of birds and mammals in the United States, poison ivy by 61, ragweed by 71, clover by 40, and cultivated corn by 190. These numbers do not indicate the use of these species by insects. One of the animals that depends on the oak is the squirrel, which also consumes insects, birds, young birds, eggs, and the fruits and seeds from 14 species of trees. It has been estimated that there is one plant species for every seven to 15 animal species. In North America alone, 3,000 of the estimated 20,000 native species are at risk of extinction.

Understanding the needs of any one species means understanding the needs of that organism throughout its life cycle. The tiger swallowtail butterfly depends on the ash tree as the host plant for its larvae and the mourning cloak butterfly depends on the elm tree. The adult butterflies require other species to supply the nectar that they need to live to reproduce. The removal of ash or elm trees removes tiger swallowtail and the mourning cloak butterflies from their ecosystems.

Food chain interactions (the transfer and loss of energy) result in a community trophic structure (energy structure). This trophic structure can be measured and graphed. The structures and graphs are called **ecological pyramids**. The ecological pyramid is based on the trophic classification. (See Figure 2-19.)

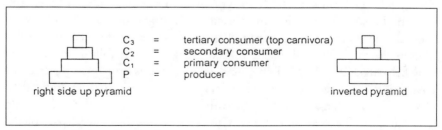

C_3 = tertiary consumer (top carnivora)
C_2 = secondary consumer
C_1 = primary consumer
P = producer

right side up pyramid   inverted pyramid

Figure 2-19: Ecological Pyramid Structure

Producers form the base of ecological pyramids. The pyramid can be either **right side up** (the producer base is larger then the next higher level) or **inverted** (the producer base is smaller than the next higher level). There are three types of ecological pyramid: **numbers, biomass, energy.** When you are looking at ecological pyramids from different kinds of ecosystems, it is necessary to remember what organisms occupy each level. For example, are the producers phytoplankton, grass, or trees?

The **pyramid of numbers** represents the number of organisms at each trophic level. The shape of the pyramid differs, depending on the size of the producers in the systems. If the producers are small (e.g., grass, phytoplankton), the pyramid is right side up. If the producers are large (trees), the pyramid is inverted. (See Figure 2-20.) Numbers pyramids overemphasize the importance of small producers.

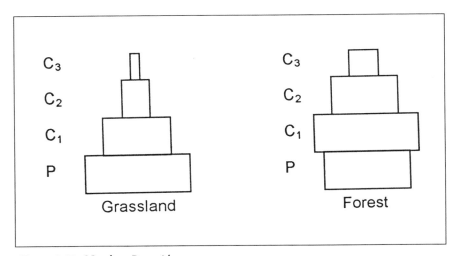

Figure 2-20: Numbers Pyramid

Biomass pyramids measure the amount of living matter at each level, and present this information as total dry weight. To find the total dry weight, you must collect the matter, dry it out, and weigh it. You can only measure the amount present at one point in time. This pyramid can also be either right side up or inverted. If the producers are small, the pyramid will be inverted. If the producers are large, the pyramid will be right side up. (See Figure 2-21.)

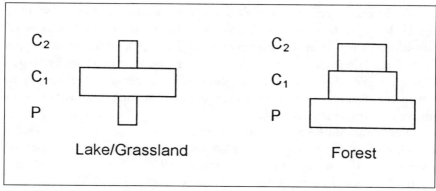

Figure 2-21: Biomass Pyramids

The biomass pyramid overemphasizes the importance of large producers. This pyramid does not give an accurate picture of the productivity of the system because it does not take into account the harvesting of the producers. Because in lakes and grasslands the producers are small (phytoplankton and grass) and are continually harvested, the biomass is small.

The **energy pyramid** measures the rate of energy flow or the productivity at each level. Since energy is lost with each transfer, energy pyramids are always right side up. (See Figure 2-22.)

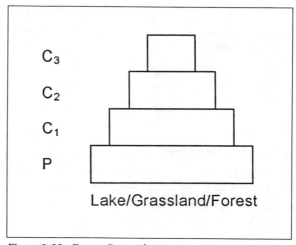

Figure 2-22: Energy Pyramid

All energy pyramids are right side up and can easily be used to compare different communities. Energy pyramids give a functional view of the community. In every ecosystem, there is less and less energy available as you go farther and farther from the producer level. This has tremendous application to the world food problem. Organisms at the C_3 level get the least energy. To get the most energy, you need to be at the C_1 level. You get more usable energy from an acre of corn than from an acre of beef. Therefore, in order to feed more people, consumption patterns must change. The consumption of meat will have to decrease as the grain necessary to feed animals will be needed for human consumption. (See Figure 2-23.)

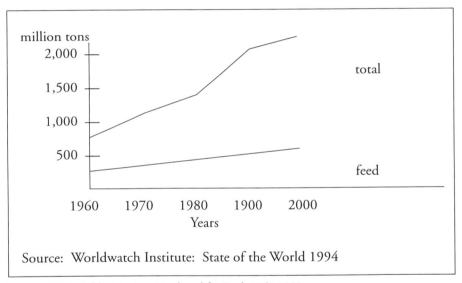

Source: Worldwatch Institute: State of the World 1994

Figure 2-23: World Grain Use, Total, and for Feed, 1960-1992

The following are general statements about energy flow in biosphere:

1. The majority of organisms ultimately depend on sunlight for their energy.
2. Energy is unidirectional through a system; it does not cycle.
3. Energy transfers through a food chain are progressively smaller as heat is lost at each level.
4. The amount of energy and the efficiency of its transfer vary from one system to the next.

5. There are three major kinds of organisms involved in the energy flow: producers, consumers, and decomposers.
6. There are usually two major pathways through an ecosystem: grazing and detritus.
7. In terrestrial systems, biomass generally increases.
8. Producers are materials (nutrients) limited and animals (heterotrophs) are energy limited.

Energy flows unidirectionally through ecosystems and most nutrients cycle within a system (e.g., iron, calcium). There are substances that act differently, however. These substances accumulate though a system. The materials increase in concentration as they progress through a food chain. This is called **food chain concentration**, or **biological magnification**. These materials can occur naturally or be man-made. (See Table 2-11 for examples.)

Table 2-11: Materials that Exhibit Biological Magnification, Their Sources and Effects

Material	Source	Effect
pesticides (DDT, PCB, etc)	man-made	hormone abnormalities, birth defects, cancer
heavy metals lead mercury	industrial processes paint, gasoline industrial waste	neurological damage, genetic damage neurological damage
radioactive materials (strontium 90, etc.)	above-ground nuclear testing, power plants	genetic damage cancer

Many of these materials are water soluble and are stored in the fat of organisms. As more and more of these organisms are consumed, the concentration of stored material becomes greater.

One of the best-documented examples of biological magnification is the pesticide DDT. DDT was widely used in the United States to control mosquito populations and agricultural pests. (See Table 2-12.)

Table 2-12: DDT as an Example of Biological Magnification

Food Chain Level	Organism Source	Concentration of DDT parts per million (PPM) of water
0	water	0.00005
first	plankton	0.04
second	minnow	0.94
third	needlefish	2.07
fourth	sea gull	6.00
fifth	osprey eggs cormorant	13.80 26.40

Table 2-12 shows how the amount of DDT increases in organisms at different levels of the food chain. DDT was sprayed on the water in Long Island, New York, in concentrations well below those set by government agencies. It was thought that DDT would control the insect population, then be dispersed by the wave action. But DDT was absorbed by the detritus and stayed in the food chain.

The increased DDT concentration did not directly kill the plankton, minnows, needlefish, gulls, ospreys, or cormorants, but at the highest DDT concentrations, mature adult organisms were affected. DDT was not lethal to the individual, but it was lethal to the species. How?

As the level of DDT increased, the reproductive rate of the top predators (e.g., the osprey and the eagle) declined until both species were in danger of becoming extinct. When the DDT concentration in adults rose, hormone systems were affected and egg-shell formation was damaged. The amount of calcium in egg shells fell, egg shells became so thin that the weight of adults birds caused them to crack. DDT concentration was also high enough to kill developing embryos. Thus, DDT was lethal to the species by killing off the developing embryos and by decreasing reproductive success, but it was not lethal to adults. DDT, as man-made substance, is currently banned from use in the United States because of its effect on living organisms. The populations of ospreys and eagles in the United States have increased since the ban took effect and the amount of DDT in the environment has degraded, losing its toxicity. However, DDT is still widely used as a pesticide in other countries. DDT is manufactured in the United States and sold to other countries.

Not all of the materials that increase through the food chain have the same effect as DDT, but they all have some deleterious effects on living organisms. Over the years, the amount of toxic material in the environment has steadily increased and now threatens whole systems. We are most aware of this when we hear that some swordfish have such high concentrations of tin that they cannot be consumed, that some tuna have mercury levels that are too high for consumption, that fish cannot be eaten from native rivers because of concentrations of PCBs, and that milk must be dumped because its strontium 90 levels are too high. The mechanism of biological magnification of toxic materials directly affects the world food supply, the functioning of ecosystems, and world health.

Another example of DDT use illustrates the interactions between biological magnification, food webs, and the precautionary principle. In Borneo, in Malaysia, malaria was a chronic problem and 90% of the population had the disease. The World Health Organization (WHO) decided to eliminate the problem by exterminating the mosquito that transmitted the disease. Two pesticides, DDT and dieldrin, were sprayed inside the huts of the people. This practice did, in fact, eradicate the mosquito, and the malaria problem was brought under control. However, there were other effects of the spraying. The thatched roofs of the huts began to collapse and the number of rats in the villages began to increase. Why?

It turned out that the DDT not only killed the offending mosquitoes, but it also killed a parasitic wasp. The wasp parasitized moth larvae that fed on the thatch. With the extermination of the parasitic wasp, the moth larvae population increased enough to destroy the thatched roofs. A second result of the DDT spaying was that the rat population increased rapidly, and with it the potential of other diseases (e.g., plague, typhus). The DDT contaminated the cockroaches that lived in the huts. These cockroaches were eaten by gecko lizards, which were, in turn, eaten by the village cats. Due to the biological magnification of the DDT, the cats were poisoned. The decline of the cat population meant that the rat population was no longer under control. The rise in the rat population threatened a health problem.

Both the roofing problem and the rat problem were easy solved: the first by putting on new roofs and the second by importing more cats. The cats were actually parachuted into the villages by the Royal Air Force in "Operation

Cat Drop." The story sounds funny because the problem could be easily fixed, but there is a very important lesson to be learned. This example illustrates how little we really understand ecosystem interactions, how using technology can cause unexpected problems, and the importance of using the precautionary principle on a daily basis.

Most environmental problems cannot be solved so easily and so cheaply in terms of biodiversity, human life, and money. It is, therefore, important to limit environmental damage by paying close attention to the complex interactions of systems and to the precautionary principle.

Soil

Soil is the surface layer of the earth's crust. It is the product of the weathering of the parent material, the decomposition of organic matter, and the activities of the biotic community. Soils differ from ecosystem to ecosystem. To understand an ecosystem, you must understand the soil, because it directly determines which plants will be present and indirectly determines which animals will be present. Plants need nutrients and water to grow, and the soil determines which nutrients will be present and how much water will be available to the plants. Soils have physical composition, physical structure, chemical composition, chemical processes, a biotic community, and biotic interactions. All of these things determine the soil type.

The weathering, or breaking down, of the parent material produces pieces of different sizes varying from gravel (largest) to sand to silt to clay (smallest). Different soils have different amounts of these particle sizes. Soil composition is important because it determines the permeability of the soil, or how fast water can move though the soil. The particles of the soil clump together and can trap some of the water. Trapped water is called soil water, or capillary water. Permeability determines how much and how long water will be available to the plants. Water moves quickly through gravel and very slowly though clay.

As the parent material breaks down and is mixed with decaying organic matter, layers (horizons) are formed. A cross-section of the horizons, or sequence of layers, is called a soil profile. (See Figure 2-24.)

Horizon	Name	Characteristics
A	topsoil	nutrient rich with fine organic matter mixed with gravel, sand, silt, and clay
B	subsoil	accumulated material leached from the A horizon
C	parent material	partially modified parent material
R	bedrock	unmodified parent material

Figure 2-24: Soil Profile

A litter layer (undecomposed dead organic matter) lies on top of the A horizon (e.g., leaves). How quickly the parent material breaks down and the soil profile develops depends on many factors, including type of parent material, climate, topography of the area, and the biotic community. Different soils develop at different rates of speed. The profile is an expression of the physical structure of the soil.

The chemical composition, or which nutrients are present in the soil, depends on the type of parent material and the kind and amount of organic material that is decomposed. The chemical composition is important because chemicals are the nutrients that plants need to live (e.g., nitrogen, iron, phosphorus, potassium). Plants take in nutrients as dissolved ions and molecules through their root systems. Because most of the nutrients are dissolved in water, the amount of water in the soil is important. Plants also need gasses, such as oxygen and nitrogen, which are found in the air spaces between the soil clumps or dissolved in the water. Some minerals are needed in large amounts (**macronutrients**) (e.g., nitrogen, phosphorous, potassium), and others are needed in small amounts (**micronutrients**) (e.g., iron, sulfur, zinc).

Different soils have different chemicals available in differing amounts. One important chemical characteristic of soil is the amount of hydrogen ions, or the pH. The soil's pH determines which minerals are available and in what form they are available. If you change the pH of the soil, you can alter its chemical processes and make some nutrients inaccessible to the plants. For example, in soils with a pH of 8, calcium is available for plant use, but iron is not.

The biotic soil community is essential for degrading dead organic matter and returning the nutrients in their elemental state to the soil. Most of the soil community is microscopic (e.g., bacteria, fungi). Even the larger soil organisms, such as slugs, worms, beetles, mites, spring tails, various kinds of larvae, and sowbugs, are usually not seen above ground. Although most of the work of this community goes unobserved by the human eye, the activities of the ecosystem would cease without the soil community because nutrients would not be recycled thorough the system. The interaction of the biotic community is essential to maintain a healthy soil and, therefore, a healthy ecosystem. These organisms are part of the detritus pathway.

Soils can be classified as: **young, mature,** and **old.** A **young** soil accumulates organic matter faster than it is lost, and therefore continues to develop a profile. A **mature** soil gains and loses soil in equal amounts, so the profile stays the same. An **old** soil loses material faster than it gains material, and nutrients are leached away faster than they can be replaced, so the A horizon decreases. Young soils are best for agriculture because they are the most productive. Only 24% of the land area of the world is suitable for intensive agricultural use (i.e., has young soil, good climate, and sufficient water).

Soil types are unevenly distributed around the world. One reason that the United States can produce so much food is that it has a great deal of young soil. Of the productive 24% of world's soils, the United States has 38%. One way to increase soil productivity is to increase the amount of available nutrients by adding fertilizers and by increasing the amount of water by irrigation.

Different soils require different methodologies for productive agriculture. The soil of the tropical rainforest is nutrient poor because most of its nutrients are locked up in the above-ground biomass (plants). This soil compacts and erodes easily. High temperatures coupled with seasonal, torrential rains leach away any nutrients present in the soil.

When the agricultural methods of the developed world were imported to recently logged tropical rainforest areas to increase food production, they resulted in disaster. Without the forest to hold it in place, the soil washed away. Heavy machinery compacted what soil remained, making it unusable. Because, as mentioned earlier, tropical rainforest soils are nutrient poor, they can produce food for only a few years. Thus, the use of tropical rainforest soils for Western-type agriculture demonstrates the importance of the precautionary principle. You must understand what you are doing, where you are doing it, and the possible outcomes *before* you tamper with an ecosystem.

It is possible to produce food in a tropical rainforest, but the agricultural methods used must rely on more traditional farming methods. The use of C_4 plants, nitrogen fixing plants, multiple cropping, adding fertilizer to the soil, doing less or no plowing (no heavy machinery), and planting perennial crops within the forest itself would allow reasonable food production in these areas.

Soil Erosion

Soil is naturally lost to ecosystems by wind and water action. Soil loss is called **erosion**. Natural erosion occurs at a slow rate and the lost soil is replaced by the action of the biotic community and weathering. If soil loss occurs faster than it accumulates, however, the productivity of the soil decreases. Farms are not natural ecosystems, so they cause damage to the larger ecosystem. Crops take nutrients out of the soil, and when they are harvested, those nutrients are removed permanently. Unless these nutrients are replaced, the soil will become nutrient poor. For example, one ton of wheat removes 18.2 kg. of nitrogen, 3.6 kg. of phosphorus, and 4.1 kg. of potassium from the soil. Heavy machinery for plowing, cultivating, and harvesting damages the soil structure, and pesticides and herbicides damage the biotic community of the soil. These human activities degrade the fertility of the soil and accelerate soil erosions.

The United States experienced massive soil loss in 1934 during the Dust Bowl. The tremendous loss of topsoil was caused by drought conditions coupled with poor agricultural practices. After the Dust Bowl, U.S. farmers began using agricultural methods that would result in increased soil accumulation. These methods included: crop rotation, buffer zones, terracing, and contour farming. Farming the same crop season after season robs the soil of nutrients and encourages pest problems. **Crop rotation** returns nutrients to the soil. By planting legume plants (plants in the pea family), nitrogen lost to the harvested crop can be replaced and soil fertility increased. The planting of forest **buffer strips** (zones) reduces wind speed and soil removal. The trees decrease the velocity of the wind and silt is dropped. **Terracing** is the cutting of a slope by creating a series of horizontal steps so that, when water run-off occurs, soil is dropped or deposited on the next lower level. **Contour farming** is plowing that follows the shape of the land, which results in reduced wind erosion. The above agricultural practices improved the quality of U.S. soils from 1935 to the 1970s.

Starting in the 1970s, soil losses in the United States began to increase and, consequently, food productivity declined. Current soil loss can be attributed to two trends: **agricultural industrialization** and increased **urban sprawl**. Agriculture has changed from a family enterprise to big business. Farming is now thought of as cash crop production. The industrialization of farming is called **agribusiness**. To make the most money per acre, you need to increase productivity and decrease production costs.

How do you do this? First, you grow only one crop (**monoculture**) on vast amounts of acreage. This means no crop rotation and no period of recuperation to allow the land to lie fallow. If you plant one kind of crop, you can limit the amount of specialized machinery you need. With huge acreage, you can plow, cultivate, and harvest without stopping for farm buildings, buffer zones, etc. You use high-yield crops, which require chemical fertilizers for increased soil nutrients, herbicides to cut down on plant competition, and pesticides to cut down on insect competition. You use heavy machinery because it is cheaper than manual labor. These practices have increased crop yield at the long-term expense of soil fertility, soil productivity, and increased soil loss.

As human populations increase and more people move into the areas around cities, the amount of **urban sprawl** increases. More homes, roads, and businesses require more and more land surface. As the land is covered with concrete, asphalt, and buildings, it is lost to the functioning ecosystems. Much of this urban development has been in agricultural areas, therefore, food production has decreased. As areas undergo urban development, topsoil is lost to erosion. It is carried into neighboring streams, lakes, and rivers, and blown away by the wind.

Without a vegetative ground cover, soil is vulnerable to erosion. The roots of the plants bind soil in place. Anything that removes plants from an area removes the roots that hold the soil. Plants have two types of root systems: **fibrous** and **tap**. A **fibrous** root system has many roots of equal size close to the surface of the soil (e.g., grass). You'll notice when you pull out a grass plant that you also remove a clump of attached soil. Fibrous roots can capture water as it moves through the soil. A **taproot** system has a single major root with a few smaller branch roots. The taproot goes deep into the soil to reach water in lower water tables (e.g., a dandelion).

Different soil types can tolerate different amounts of annual soil loss without affecting the soil's productivity. A deep soil (at least six inches of topsoil), under normal conditions, can tolerate a loss of five tons of soil per acre per year and a poor soil two tons per acre per year without damaging productivity. However, it is estimated that currently 10 to 20 tons of soil are lost per acre per year through erosion. The loss of 10 tons per acre per year is equal to about one inch of topsoil. For every inch of topsoil lost and not replaced, crop yield decreases a minimum of 10%. Under good conditions, it takes about 17 years to produce an inch of topsoil. It is estimated that, in the United States, the average soil loss to erosion occurs seven times faster than it is formed, and that 23 billion tons of soil are lost each year worldwide from croplands.

To increase crop yields, soil erosion must be reduced and soil quality improved. Agricultural and urban development practices will have to change. Agricultural soil loss can be reduced by changing plowing and cultivation methods. By reducing the amount of plowing and by keeping a continual vegetation cover (**no tillage**), or by keeping the soil covered as much of the time as possible (**conservation tillage**), there will be less erosion by wind and water.

However, with less cultivation, the farmer faces increased weed problems and often turns to herbicides. Soil fertility can be increased by using crop rotation and by decreasing the use of chemical herbicides and pesticides. The loss of soil to urban sprawl can be reduced by well-planned urban development. Planned urban growth and not random, opportunistic development is needed. How many shopping malls and discount stores does any area really need? Well-developed plans would cut down on the loss of agricultural area and soils.

Land degradation is worldwide and directly affects productivity. Several other human activities contribute to soil loss: poor lumbering techniques, mining, overgrazing that results in desertification, and over-irrigating desert soils (salinization). (See section on biospheric problems.) (See Table 2-13.)

Nutrients (Biogeochemical Cycles)

Biogeochemical cycles are the pathways by which chemical elements move from the abiotic environment to biotic environment and back into the abiotic environment ($^{biotic}_{abiotic} \rightleftarrows$). The word *biogeochemical* can be broken down in this way: *bio* = living, *geo* = geological (earth, water, air), *chem* = chemical forms, or combinations of elements and processes. In other words, as elements move through the ecosystem from the nonliving environment to the living community and back to the nonliving environment, they exist in different chemical forms or combinations. Unlike energy, which flows unidirectionally into and out of the system, elements are used repeatedly.

Table 2-13: Worldwide Land Degradation Due to Human Activity, 1945-1994

Region	Over-Grazing	Deforestation	Agricultural Mismanagement	Other*	Total	Degraded Area as Share of Total Vegetated Land
	(millions of hectares)					(percent)
Asia	197	298	204	47	746	20
Africa	243	67	121	63	494	22
South America	68	100	64	12	244	14
Europe	50	84	64	22	220	23
North & Central America	38	18	91	11	158	8
Oceania	83	12	8	0	103	13
World	679	579	552	155	1,965	17

*Other includes domestic use and bioindustrial activities, such as pollution.
Source: Worldwatch Institute: State of the World 1994.

Because elements move around and around, they are said to *cycle* through the system. There are 110+ chemical elements in the biosphere, and approximately 30 of these are required for living organisms to function properly. These essential elements are called **nutrients**. Nutrients required in large amounts are called **macronutrients**, those required in smaller amounts are called **micronutrients**, and those needed in infinitesimal amounts are called **trace elements**. These chemical elements are unevenly distributed throughout the biosphere and can be found in different chemical forms in different areas. The circular pathways of essential nutrients are called **nutrient cycles**.

All life is composed of elements in various chemical forms or combinations. Elements taken in by living organisms and organized into organic compounds, which build organisms and keep them running. All living organisms are composed of four major kinds of organic compounds: carbohydrates (sugars and starches), proteins (enzymes, etc.), fats (lipids, hormones, etc.), and nucleic acids (DNA, RNA, etc.). Without the right kinds and amounts of nutrients, living organisms cannot function properly. For example, if you do not get enough iron to make red blood cells, you become anemic. If a green plant does not get enough magnesium, it cannot produce enough chlorophyll for photosynthesis.

Each cycle has two areas (**pools**) where the element is located: the **reservoir pool** and the **exchange (active, cycling) pool**. The **reservoir pool** contains a large amount of an element, which moves out of the pool slowly. The element is either in a form that is chemically unusable by the biotic part of the system, or it is physically remote (e.g., locked into the parent material). It is nonbiological and the element is relatively unavailable to the biotic part of the system. The **exchange (active, cycling) pool** contains a smaller amount of an element that moves rapidly between the biotic and the abiotic parts of the ecosystem. The element is in a form that is usable by the organism or is exchangeable. Nutrient cycles are energy driven.

Nutrient cycles are classified by the location of the reservoir pool. A **gaseous cycle** has its reservoir pool located in either the atmosphere or hydrosphere. Gaseous cycles, because of their ability to adjust quickly to changes in the system, are called *perfect* cycles (e.g., carbon, oxygen, nitrogen, water). However, you will see that there are limits to the self-adjusting abilities of gaseous cycles. They are not "perfect" in the sense that they cannot adjust to any magnitude of disruption.

A **sedimentary cycle** has its reservoir pool located in the earth's crust. Sedimentary cycles are less "perfect" because their elements are relatively inactive and unavailable (e.g., phosphorus, iron, potassium, sulfur). This makes the system easy to disrupt. What is important in a cycle is not how much (quantity) of the element is in each of the pools, but how fast (rate) the element is released into the system.

The **rate of exchange**, or transfer, between the pools more clearly determines the structure and function of a system than does the quantity of the elements present. The rate of exchange is called the **turnover rate**. If 10 units of the element are released each hour and the compartment holds 1,000 units, the turnover rate is 1%. The **turnover time** is the time required to replace the total amount of the element in the compartment. If the element is replaced 10 units each hour and the compartment holds 1,000 units, the turnover time is 100 hours.

The rate of uptake by the biotic community and the rate of release back into the environment can vary seasonally. The uptake of carbon, as carbon dioxide, is greater in the spring and summer as deciduous trees grow leaves and carry on photosynthesis. The release of CO_2 to the atmosphere is greater in the fall when the leaves fall off the trees and decompose. (See the section on the carbon cycle.) How long an element stays in a compartment of a cycle is called its **residence time**. The amount of time CO_2 stays in the atmosphere is five to seven years. How much and how fast elements move through systems is relatively unknown for most systems.

The location of different elements or nutrients can vary from system to system. (See Table 2-14.) The cycles in the tropical rainforest are primarily biological, and those in the temperate forest are primarily physical. A clear understanding of nutrient cycles is essential when working with different ecosystems. Because element cycles are most often discussed for local ecosystems, it is easy to forget that elements and materials circulate on a global scale. A farmer in Iowa is more likely to be concerned about the amount of nitrogen available in the soil on his farm, not in the amount available in France. However, materials do not remain in local areas. Materials move around the biosphere in **global circulation patterns**. What is dumped into the soil, water, or air in your hometown moves around the world and eventually comes back to you.

Table 2-14: Nutrient Distribution in Temperate and Tropical Rainforests

	Tropical Rain	Temperate
carbon in vegetation	75%	50%
carbon in litter and soil	25%	50%
nitrogen in biomass	50%	6%
nitrogen in biomass above ground	44%	3%

The inert gas argon can illustrate global circulation. Argon is a naturally occurring gas that makes up about 1% of the atmosphere and does not combine with other elements. With each breath, you inhale and exhale millions of argon atoms. If you live to be 60 years old, you will breathe in at least one argon atom that has been breathed by every living thing that is alive today and has ever been alive on earth. Another example of global movement is that today you will breathe in 17 atoms of argon that you breathed in one year ago today, no matter where you were a year ago, be it Alaska or China. Materials move globally.

You may think that global circulation doesn't affect you, but that's not so. Let's return to the DDT example. DDT is a man-made pesticide known to cause environmental damage. DDT is now found in the ice of Antarctica and in the animals that live there. DDT has never been used in Antarctica, so how did it get there? It has been carried around the earth by global circulation. Even though DDT is no longer used in the United States because it is unsafe, there are companies in the United States that manufacture it and sell it to other countries. It does make a different what everyone on earth does because everyone's actions affect all the other systems. What goes around, comes around.

All biogeochemical cycles have the same basic structure. (See Figure 2-25.) How the components of a cycle work varies from cycle to cycle. The cycles do not work independent of the rest of the ecosystem; they are interconnected. The functioning of each cycle depends on a healthy community of decomposers. Anything that alters the environment of the decomposers will affect the recycling of nutrients.

I am going to discuss briefly six of the 30 essential nutrient cycles to give you an idea of their biological importance, how they work, and how they have been disrupted by human activities. When you study a cycle there are questions you should ask:

1. Why is the element important to living organisms?
2. Where, globally, is the active pool located?
3. Where, globally, is the reservoir pool located?
4. What type of cycle is it?
5. What form of the element is most abundant?
6. How is the element incorporated by autotrophs?
7. How is the element incorporated by heterotrophs?
8. How is the element returned to the abiotic part of the cycle?
9. Which part of the cycle is most vulnerable to disturbance?

10. How does more of the element get into the cycle? (What is the global natural uphill gain)?
11. How does some of the element leave the cycle? (What is the global natural downhill loss)?
12. What are the effects of natural events (e.g., seasons, weather) on the cycle?
13. Are some organisms more important than others in maintaining the cycle?
14. How does the cycle interact with and affect other cycles?
15. What human activities have affected the uphill gain?
16. What human activities have affected the downhill loss?
17. How has human activity affected the cycle?
18. How do changes in the cycle affect aquatic ecosystems?
19. How do changes in the cycle affect terrestrial ecosystems?
20. What can be done to decrease the human effect on the cycle?
21. What can *you* do to decrease human effect on the cycle?

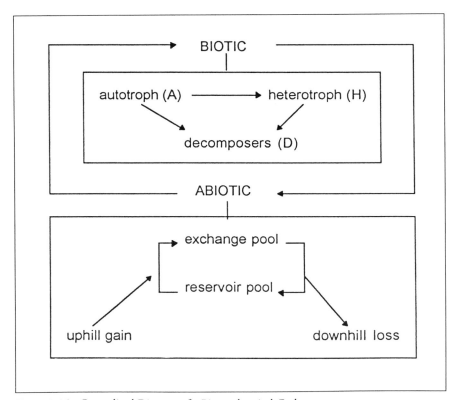

Figure 2-25: Generalized Diagram of a Biogeochemical Cycle

Carbon Cycle

The element carbon is found in all organic molecules and is the chemical skeleton of the four major classes of organic compounds: fats, proteins, carbohydrates, and nucleic acids. The carbon cycle is a gaseous cycle with a small active atmospheric exchange pool of carbon dioxide (CO_2), a large aquatic reservoir pool, and a secondary reservoir in sedimentary rocks. Carbon is moved through the cycle by autotrophs, who fix carbon dioxide into organic compounds by the process of photosynthesis. The organic compounds are then passed on to heterotrophic organisms and decomposers through the grazing and detritus food chains. Carbon is returned as carbon dioxide to the atmosphere through cellular respiration.

There is a natural uphill gain to the cycle caused by volcanic activity. Natural downhill losses occur in two ways:

1. when organic matter is trapped in sediments and accumulates as fossil deposits, and
2. when carbon dioxide is lost in the formation of aquatic sediments.

When carbon dioxide dissolves in water, it forms carbonic acid, which, through a series of chemical steps, combines with the calcium to form a bicarbonate (Ca_2CO_3). Bicarbonates are not very soluble. They precipitate out of solution and sink to the bottom of a lake or ocean to form sediments. These chemical reactions are reversible. If the amount of carbon dioxide decreases in the aquatic system, the bicarbonates break down to release CO_2. The oceans act as a buffer for the amount of global atmospheric carbon dioxide. Carbon dioxide resides in the atmosphere from five to 10 years before returning to the ocean. (See Figure 2-26.) Forests are the primary reservoir of biologically fixed carbon. The concentration of carbon dioxide in the atmosphere fluctuates throughout the year. There is more CO_2 in the atmosphere in the fall and winter because of the increased decomposition of vegetation. There is less CO_2 in the spring and summer because more photosynthesis is occurring. The amount of photosynthesis and cellular respiration balances out over the year.

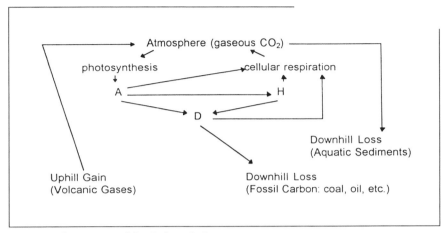

Figure 2-26: Diagram of the Carbon Cycle

Human activity has disrupted the natural carbon cycle by increasing the amount of carbon dioxide in the atmosphere. The amount of CO_2 has been increasing since the start of the Industrial Revolution in the 1750s. (See Figure 2-27.) There are primarily two human activities that have contributed to the increase in CO_2: deforestation and the burning of fossil fuels. Deforestation is the removal of forests. Several activities, such as urban expansion, agricultural usage, grazing usage, and pulp production, have contributed to deforestation, but its primary cause has been harvesting for wood (lumber). The increase in deforestation has accelerated with the innovation of advanced lumber techniques (the use of machines). Since 1950, one-fifth of the earth's forested areas have been cleared. Since 1950, logging has doubled. Between 1970 and 1991, the total world consumption of forest materials has grown 44%.

Although you hear about the loss of the tropical rainforests because of deforestation, you should know that temperate forests also suffer from overharvesting and deforestation. Ninety percent of the old-growth forests in the northwestern United States have already been logged.

The most widely used tree harvesting method is called **clear cutting**. This is the fastest and cheapest way to harvest timber. The lumber company moves into a forested area and harvests, or cuts down, all of the trees in the area. The logs that can be used are hauled out and the remaining material (unusable timber and biomass) is burned. The deforested land is then planted with young trees. The forest is turned into a **tree farm**, or **tree plantation**. A tree farm is

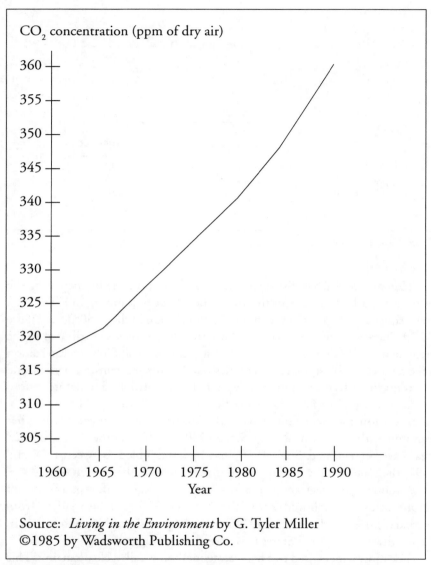

CO$_2$ concentration (ppm of dry air)

Source: *Living in the Environment* by G. Tyler Miller
©1985 by Wadsworth Publishing Co.

Figure 2-27: Monthly Average Carbon Dioxide Concentrations from 1960-1990

not a forest. A forest has biodiversity and age diversity. All of the trees in a tree farm are the same age and the same species (monoculture). Many clear cut areas have had to be replanted several times and are still not forested. With increased soil erosion, which is caused by clearing the land of the plant material that holds the soil in place, the nutrients of the area are carried away, impoverishing the soil for future vegetation. Tree plantations are cut every 60 to 70 years on a rotation pattern, so that a mature forest is never established.

The removal of the trees disrupts the carbon cycle in two ways: the amount of photosynthesis is decreased, leaving more CO_2 in the atmosphere, and the burning of the waste biomass puts CO_2 into the air. It is important to note that deforestation does not only disrupt the carbon cycle; it has other long-term ecological effects. Mature forests do more than help maintain the carbon cycle. They also help to control climate, maintain biological diversity, retain water in the water cycle, keep soils from eroding, provide habitat for other living organisms, and keep the nutrients in the area. Clear cutting disrupts all of these functions.

You can see how deforestation affects other organisms by examining its effect on the salmon populations of the northwestern United States. With clear cutting, forest soil erosion has increased and silt has washed into the streams where salmon spawn (lay their eggs). The eggs are covered with soil (silted over) and smothered. Removing the trees also means removing the shade, which causes the water temperature to rise, stressing the young fish in the river. Thus, deforestation has had a tremendous negative impact on salmon fisheries.

There are other methods for harvesting trees. **Selective cutting** harvests only good lumber trees. **Sustainable yield** forestry practices require that timber cutting is limited to the amount that can be replaced by the forest in one year. These methods make lumber and wood product production more expensive, but they help to maintain a functioning mature forest in the ecosystem. We are unable to replant a forest in all of its complexity and reestablish all of its functions in the biosphere.

The cutting of forests to provide grazing pasture for beef has also encouraged deforestation. Because of the kinds of soils found in the tropics, the pasture is only good for feeding beef for about three years, then it must be abandoned.

What can be done to decrease deforestation? The answer is to cut fewer trees. This can be accomplished by finding material alternatives to wood, recycling, and reusing the lumber we now have. Houses do not have to be made out of wood products. In many parts of the world, other building materials are used. In the southwestern United States, for example, the following new technologies (using recycled materials) and revived old building technologies are being used to construct houses: adobe, rammed earth, and straw bale.

Much of the timber cut to produce wood pulp is used to produce paper products. (See Figure 2-28.) Pulp and paper product production are among the biggest polluters. Not only do they use large amounts of electricity and water, but they are also major producers of toxic water pollutants. Paper does not have to be made from wood pulp. "Treeless paper" can be made from bamboo, jute, hemp, and kenaf. These plants grow rapidly on marginal land. For example, kenaf produces two to four times more pulp per hectare than does the more commonly used pine.

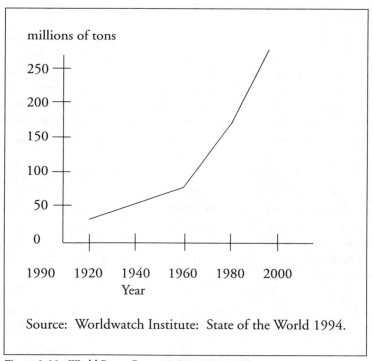

Figure 2-28: World Paper Consumption, 1931-1991

Another way to decrease deforestation is to decrease consumption. (See Figure 2-29.) We can also recycle wood and paper products as much as possible and price wood products at what is called the **environmental price**. **Environmental pricing** means that the price of the product should reflect the price of correcting the environment damage, or lost ecological function, done when harvesting and processing the material. Environmental pricing takes into account how much ecological function has been lost in the production of the product.

Currently, production cost is determined by a nature-blind pricing system. The full value price of a hamburger that is produced from beef raised on pasture made from a cleared rainforest, for example, is approximately $200. As prices go up, usage goes down. The quality of the product and its value become more important than quantity and volume. Consumption goes down as prices go up. The way to decrease deforestation caused by grazing demand is to change the human diet from large amounts of meat to a more vegetarian diet.

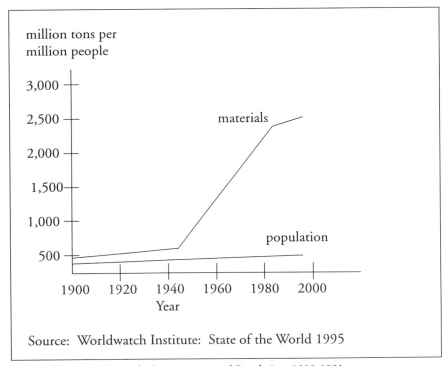

Figure 2-29: U.S. Materials Consumption and Population, 1900-1991

The other major disruption to the carbon cycle is the burning of fossils fuels (e.g., coal, oil, natural gas). When these fuels are used directly or indirectly to produce products, such as gasoline and electricity, to produce energy, the carbon stored in these materials is returned to the atmosphere. These sources of energy produce massive amount of CO_2. (See Table 2-15.)

Table 2-15: Pounds of CO_2 Produced Based on the Types of Fuel

Carbon Dioxide Production

Fuel Type	Units	Conversion Factor to Pounds of CO_2 Produced
Electricity	kilowatt (kwh)	1.8
Natural Gas	therm	12.0
Gasoline	gallon (gal)	19.0
Wood	pound (lb)	1.9

Why does it make a difference if the CO_2 concentration in the atmosphere increases? Atmospheric CO_2 allows shorter wavelength energy (sunlight) to pass through the atmosphere, where it is absorbed, then radiated back into space as heat energy (long infrared wavelengths). CO_2 absorbs (traps) the long wavelengths and reradiates them back to earth, warming both earth's surface and the lower atmosphere. This naturally occurring phenomenon is called the **greenhouse effect**.

The temperature of the earth changes on a 200,000-year cycle from warm to cold, with a change of +/- 7 to 8 degrees F. This natural fluctuation results from a change in the shape of the earth's orbit around the sun and the tilt of its axis. When the earth is farthest away from the sun and tilted away from the sun, there is an ice age. Between ice ages, the greenhouse effect warms and cools the earth. When there are more greenhouse gases in the atmosphere, the temperature goes up. When there are fewer gases, the temperature falls.

The phytoplankton in the ocean regulate the amount of carbon dioxide in the air. As the global temperature rises, phytoplankton populations increase, carry on more photosynthesis, and take up more CO_2. As the amount of CO_2

decreases, the climate cools and the algae start to die off, which means that less photosynthesis takes place, more CO_2 stays in the air, and temperatures rise again. There is an annual change in CO_2 because of the amount of photosynthesis occurring in terrestrial ecosystems. There is more CO_2 taken up in the spring and summer, and more CO_2 released by decomposition in the fall.

We can trace this natural fluctuation historically and see its direct effect on human civilization. For example, 50,000 years ago, the global temperature fell and the Bering Strait froze, creating the Bering Land Bridge, which allowed organisms to cross from Asia to North America and migrate into South America. When the temperature rose, the ice melted, covering the Bering Land Bridge and stopping the migrations.

Many cultural changes can be traced to changes in global temperature. Here are a few: In 3,000 B.C., an increase in temperature resulted in droughts that led to Egypt's rise to power. In 1,200 B.C, a decrease of eight degrees resulted in the disappearance of the Mycenaean culture. In 500 A.D., the temperature fell seven degrees, changing the weather pattern so that the nomadic peoples could no longer grow food, causing them to migrate swiftly west into the Roman Empire. Thus, a weather change was responsible for the fall of the Roman Empire. In about 972 A.D., an increase of eight degrees allowed the Vikings to explore and colonize Greenland.

Why does global temperature change affect human populations? As global temperature changes, weather patterns shift, so that the amount of precipitation and its duration and location are altered for decades. These temperature changes affect all living organisms, not just humans. Temperature is a limiting factor for all life.

How does an increase of CO_2 influence the greenhouse effect? When **greenhouse gases** reach a level in excess of what can be removed from the atmosphere, **global warming** occurs. Global warming is a human-enhanced greenhouse effect. It is the result of an unnatural rise in the earth's temperature because of increases in greenhouse gases. There are four greenhouse gases, all of which are on the rise in the atmosphere today as a result of human activity. (See Table 2-16.)

Table 2-16: Important Greenhouse Gases

Compound	Source	Contribution to Increased Greenhouse Effect
carbon dioxide (CO_2)	burning of fossil fuel, deforestation	60%
methane (CH_4)	cows, termites, decomposing garbage, wetlands	15%
nitrous oxides (No_x)	industry, fertilizers	5%
chloroflurocarbons (CFCs)	air conditioners, refrigerators, aerosols foam packaging	12%
ozone (O_3) (in the troposphere)	conversion of auto-mobile pollutants by the sun	8%

Human activity first started to influence global climate in the 1750s, with the start of the Industrial Revolution and the dependence on fossil fuels. In the 1900s, two major cultural changes continued the increase in greenhouse gases we see today: industrial mass production and the concept of consumerism.

Until this time, excess CO_2 had been removed from the atmosphere by the ocean's **Great Conveyor** and **Atlantic Engine**. The ocean runs the climate through the movement of water (ocean currents) around the globe. The water in the ocean circulates in a pathway from the Atlantic to the Pacific Oceans called the **Great Conveyor**. (See Figure 2-30.) The Gulf Stream contains warmer water that moves up the coast of South America to Central America and crosses to North Africa and Europe. When it reaches the North Atlantic, the water sinks to the bottom, where it moves along around Australia and comes up to the surface in the North Pacific, then travels back to South America. The water warms up in the tropics, and because warmer water is lighter than cold water, the Gulf Stream stays on top of the water column.

What is the **Atlantic Engine**? When the water reaches the North Atlantic, the winds from Iceland evaporate the water, which becomes saltier and heavier and sinks (approximately five billion gallons per second). As it sinks, it carries dissolved CO_2 with it, removing the CO_2 from the atmosphere. The Great Conveyor and the Atlantic Engine are essential to the removal of excess CO_2 and maintaining the global climate.

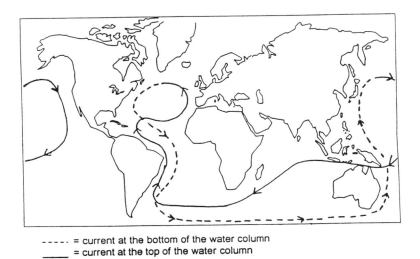

- - - - - = current at the bottom of the water column
_____ = current at the top of the water column

Figure 2-30: Diagram of the Ocean Great Conveyor

Currently, it is estimated that seven billion tons of carbon are emitted annually by the burning of fossil fuel and deforestation. Of this amount, the oceans take up approximately one- to two-billion tons, plants take up approximately two- to three-billion tons, and three billion tons are added to the atmosphere. The net result is an annual increase in the carbon load of the atmosphere. It is estimated that if the 1980 carbon levels were doubled, the resulting increase in the average global temperature would be between 2.7 and 8.1 degrees F. The temperature rise at the North Pole (between 13 and 18 degrees F) would be greater than the average global temperature rise.

What would happen if the global temperature increased? The ice caps in the North Atlantic would melt, diluting the ocean; seawater would not get saltier, so it would not sink and the Great Conveyor would slow down; and more CO_2 would remain in the atmosphere. As more CO_2 remained in the atmosphere, the temperature would continue to rise. With the rise in temperature, the phytoplankton in the oceans would begin to die, so even more CO_2 would be left in the air, and a positive feedback mechanism would be established. Excess CO_2 can only be removed by the process of photosynthesis, which means that to check global warming, we must begin massive forest planting programs and stop the deforestation that we now practice.

It is estimated that by the year 2010, there could be a rise of three degrees F, which would result in a one-inch rise in the sea level and a shift in climate patterns. This would cause a host of problems: the number and strength of tropical storms would increase, increased flooding would occur, there would be massive droughts in agricultural regions, major rivers would dry up, there would be decreases in normal food-producing areas as plant tolerance limits are exceeded, the rise of the sea level would result in the breaching of coastal sewage systems and pollution of the oceans, problems with nuclear reactors that are placed on the coasts would occur, and droughts would cause water shortages, starvation, and refugee problems.

Will global warming occur? Scientists agree that it will happen, but they do not agree on exactly when or how much warming will occur. Steps must be taken to try to limit the amount of warming. How can you do that? You can decrease the amount of greenhouse gases emitted by changing technology and social values: by controlling the fuels you use, how much you use, and by increasing your energy efficiency.

Society is fossil fuel-based. This must change. There are other energy sources: hydro, thermal, wind, and solar, to name a few. These can be used to directly make the energy needed to run our technologies. Less gas can be consumed with the new electric transportation technology and the use of mass transit systems. Changes in diets can decrease meat consumption. Fewer cows can mean less deforestation and less methane production. Electricity can be produced not by burning coal, but by utilizing hydro-, thermal-, wind-, and solar power. The consumption of material goods can be decreased.

We must, therefore, make changes in our own lives and in what we consider important. No country will be exempt from the effects of global warming, and because the developed nations are primarily responsible for the production of greenhouse gases, they must change the most. It is estimated that fossil fuel comsumption must be decreased by 60% if global warming is to be limited.

Remember that temperature is one of the most important limiting factors for all living organisms. When you change the temperature, you change the whole system: what lives where, how these organisms live, and how many can live (if at all).

Water Cycle

Water is the medium for all cellular chemical reactions and is essential to all life. Organisms use fresh water. Your body is 90% water. The water cycle, or **hydrological cycle**, is a gaseous cycle with a small active atmospheric exchange pool of water vapor, a large aquatic reservoir pool, and a secondary reservoir in the lithosphere. The water cycle is driven by solar energy through the process of evaporation.

Water evaporates from all surfaces, including soil, bodies of water, and organisms. Vegetation returns water to the atmosphere by evaporation and by **transpiration**. **Transpiration** is the process by which autotrophs move water from the soil to the leaves for photosynthesis. A mature oak tree in full leaf moves four to five tons of water from the soil into the atmosphere by transpiration every day. Water returns from the atmosphere by precipitation in the forms of rain, sleet, ice, snow, and fog. The resident time in the active atmospheric pool for water vapor is nine to 10 days. (See Figure 2-31.) The amount of evaporation and precipitation over the ocean and land is not equal. (See Table 2-17.) Excess ocean evaporation is carried by wind currents onto the land, where more precipitation occurs than evaporation. Evaporation over the oceans is critical for land systems. Of the water that falls on terrestrial systems, some is taken up immediately by vegetation, some replenishes the groundwater, some is runoff that eventually returns to the ocean, and some enters the large lithosphere reservoir pool. It is estimated that 20% of the annual rainfall is over the oceans, and that the 80% that falls on terrestrial systems is taken up by vegetation and recharges surface, ground reservoirs, and deep reservoirs.

Water that falls on land moves down by gravity (**gravitational water**) until it reaches a level at which the soil is saturated, or the spaces between the soil particles are full of water. This level of ground saturation is called the **water table**; the water found there is called the **groundwater**. Groundwater makes up 95% of the world's supply of fresh water. If the topography of the land is below the water table, streams or lakes, bogs or wetlands (**surface water**) are present.

The movement of surface water toward the sea forms patterns called **watersheds**, or **drainage basins**. When studying a system, it is important to

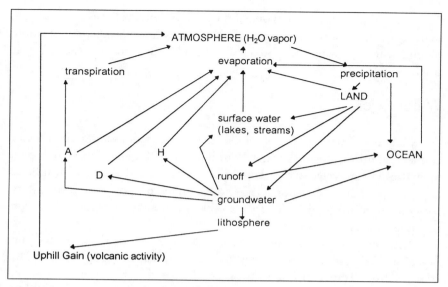

Figure 2-31: Diagram of the Hydrological Cycle

Table 2-17: Distribution of Water Movement

Area	Evaporation	Precipitation (direct return)
Ocean	10/12	9/12
Land	2/12	3/12

study the whole watershed. If you develop land by building on part of the watershed, you will affect the runoff in that area, and this will impact the rest of the watershed. If you pollute the water in part of the watershed, you may pollute all of the watershed water.

Water is constantly moving. Ground water is important because it recharges lakes and streams. Some groundwater will slowly move to deeper levels and can recharge the large pools of geologic water (**aquifers**). If the land is not flat, undergroundwater slowly moves to empty into the oceans. (To see how water is distributed worldwide, see Table 2-18.)

Surface water is not evenly distributed on the earth's surface, and usage of water varies from country to country, depending on the culture, population

Table 2-18: World Water Resources

Location	Percentage of World Supply
Oceans	97.134
Atmosphere	0.001
On Land	
ice caps	2.225
glaciers	0.015
saline lakes	0.007
freshwater lakes	0.009
rivers	0.0001
In Land	
soil moisture	0.003
groundwater	0.606

Source: *Living in the Environment* by G. Tyler Miller
© 1985 by Wadsworth Publishing Co.

size, and availability of water. Water is essential for all life on earth, and its abundance or shortage affects all societies, both now and in the future. Most of us are unaware of how much water we use daily. (See Table 2-19 for estimates of the average water use by a person in the United States in 1983.)

Every gallon of water used must be cleaned before it can be reused. The cleaning can be done naturally, but cleaning domestic and industrial water by water treatment plants takes energy. This process usually means burning fossil fuel, which results in an increase of the greenhouse gas CO_2.

Human activity has disrupted the natural water cycle by: increasing the amount of runoff through paving, ditching, draining wetlands, compacting soil, and clear-cutting forests; decreasing the large water reserves by over-pumping water for irrigation and overusing water domestically; decreasing infiltration of water into the soil by soil compaction and coverage; decreasing the amount of wetlands (which act like sponges); and polluting water sources, thereby making water unusable. Although Table 2-19 lists water usage in the United States, this is a global problem.

Table 2-19: Average U.S. Water Use in 1983

Use of Product	Average Amount Withdrawn (Gallons)
Total Use	1,953
Home Use	
Total per person (per day)	90
Drinking water (per day)	0.5
Shaving, water running (per minute)	2
Shower (per minute)	5
Tiolet (per flush)	6
Cooking (per day)	8
Washing dishes, water running (per meal)	10
Water lawn or garden (per minute)	10
Automatic dishwasher (per load)	16
Bath	36
Washing machine (per load)	60
Leaky toilet (per day)	124-24
Leaky faucet (per day)	240-48
Agricultural use (irrigation)	
Total per person (per day)	671
One Egg	40
Flour (one pound)	75
Orange	100
Glass of milk	100
Loaf of bread	150

(Continued on next page)

Table 2-19: Average U.S. Water Use in 1983

Use of Product	Average Amount Withdrawn (Gallons)
Corn (one pound)	170
Sugar from sugar beets (1 pound)	230
Rice (one pound)	560
Grain-fed beef (one pound)	800
Cotton (one pound)	2,040
Industrial and Commercial	
Total per person (per day)	1,192
Cooling water for electric power plants (per person per day)	978
Industrial mining and manufacturing (per person per day)	183
Refine 1 gallon of gasoline from crude oil	10
Steel (one pound)	35
Refine one gallon of synthetic fuel from coal	265
One Sunday newspaper	280
Synthetic rubber (one pound)	300
Aluminum (one pound)	1,000
One automobile	100,000

Source: Living in the Environment by G. Tyler Miller
©1985 by Wadsworth Pub. Co.

Increased urban sprawl, often called the "paving of America," has covered untold acres of soil with concert, asphalt, housing, and industrial buildings. To make more land available for development, wetlands have been ditched, drained, and filled. Clear-cutting forests for lumber and urban development has resulted in soil compaction and increased runoff. (See the section on the carbon cycle.)

This paving has decreased the amount of area in which water can filter through the soil, and water that does not soak into the ground runs off the land. In desert areas of the United States, such as Arizona and southern California, you see drainage systems designed to channel this runoff and avert flooding, even though these areas receive only four to eight inches of rain annually. The runoff water that would normally filter through the soil and replenish the ground water and aquifers is thus drained away. Globally, increased runoff goes to the oceans, raising the sea level and causing flooding.

The demands placed on water systems by overpopulation are especially noticeable in the arid western United States. The ever-growing population and the need for agricultural water have resulted in the damming of the great rivers. The harnessing of a river by building dams allows people to control the water for power production (**hydropower**), irrigation, and domestic/industrial use.

Hydropower is a pollution-free source of electricity, but damming a river has environmental costs. With cheap power in an area, the population density increases, thereby increasing the demand for domestic and commercial water. The Columbia River is a good example of the effects of harnessing a river. The Columbia River flows from the Canadian Rockies into the Pacific Ocean through Washington State, passing through rock, desert, and rainforests. There are over 120 dams on the Columbia River, Snake River, and their tributaries. These dams allow people to control the rate of water flow and the level of the water in the system. The changes to water flow affect all of the ecosystems that depend on the river.

These dams were built to provide water to people, which they do, but they also have had an adverse effect on commercial salmon fishing. Before the dams were built, 10 to 16 million salmon runs were common on the Columbia River system. Now the annual run is down to 300,000. Western salmon hatch in the freshwater streams of the Columbia basin, then, as hatchlings, they swim down the river to the ocean. They mature in saltwater, then return to the stream in which they were hatched to spawn.

Dams affect the salmon migration by acting as physical barriers to both the upstream migration of the adults and the downstream migration of the young fish. Dams also decrease the flow of water, which increases the time it takes for the salmon to make their journey to the ocean. Salmon change physiologically from freshwater fish to saltwater fish as they mature. If the fish do not reach saltwater by the time they have changed to a saltwater physiology, they die. The fewer fish that get to the ocean, the fewer ocean salmon can be caught. Dam building, then, has caused the salmon population to drop so low that the Northwest salmon fleet was banned from fishing in the summer of 1993. Remember that salmon are part of the river food web, and so they have become a symbol of a sick environment. These problems can be alleviated by finding a compromise between the need for hydroelectric power production and the salmon migration. If the dams are opened to allow the water to flow faster when the fish migrate downstream, the fish can get to saltwater in time. If fish ladders are built around the dams, the fish can migrate upstream. Once again, all parts of the system must be considered, not just one need.

Most organisms can only use freshwater. Pollution of natural water systems decreases the amount of freshwater available. The Great Lakes contain one-fifth of the world's surface freshwater. Water is used in industrial plants, and the Great Lakes have been a mecca for industry. The result of this industrial use and abuse of water has been pollution. There are 43 identified toxic hot spots on the Great Lakes and the rivers that empty into them. Two of the rivers that feed into the lakes, the Buffalo and Nigra Rivers, have been called "Chemical Alley" because of their high levels of pollution. Companies along the rivers and lakes dump chemical waste into the water and build toxic landfills on the land surrounding the water system. Water seeks the lowest level by gravitation, and contaminated water (polluted) seeps into the ground, entering the groundwater system, finally flowing into the rivers and lakes. Toxic chemicals such as PCBs, Myrex, and Dioxin, to name a few, have made these freshwater systems unusable.

Not only do industries pollute the water, but household products also enter the system when they are poured into sewers or household drains. Pesticides and fertilizers from the surrounding agriculture enter the system. These pollutants settle to the bottom of the aquatic system, where they undergo biological magnification through the food chain. These chemicals are toxic at low levels, causing cancer and damage to the hormone and nervous systems, and creating problems with the immune system. The effect of 80% of these chemicals is poorly understood or unknown at this time. The water may look clean, but is not.

Pollution is not limited to surface water, but it can seep into the geologic water reservoirs, the aquifers. These reservoirs are deep in the ground, where the natural cleaning processes of water movement and organism activity cannot take place. One of many people's great fears is that aquifers will become polluted and there will be no way to clean them. Seventy percent of the water used globally for irrigation comes from aquifers. A polluted aquifer directly affects the amount of agricultural production. Pollution of water systems can be direct or indirect. Air pollution can cause water pollution through toxic rain and acid rain. Both of these pollutants can destroy freshwater systems. (See limiting factors and the sulfur cycle sections.)

Wetlands areas, such as salt marshes and mangrove swamps, are found along rivers, estuaries, and ocean coasts. They are some of the most productive ecosystems in the world. They clean water, are nurseries for many ocean species, and are rich in nutrients. They act as sponges, holding water to decrease flooding in the surrounding terrestrial systems. Because they have nutrient-rich soils, many wetlands have been drained or ditched for agricultural purposes. Because they are found along the ocean, many wetlands have been prime areas for development.

One example of the devastation of wetlands is southern Florida and the Florida Everglades. The Everglades act as a water source for south Florida. They clean the water, act as a wildlife area, and carry nutrient-rich water into Florida Bay. The water in the Everglades has been diverted from its natural course for agricultural and domestic uses. As a result, 2,000 miles of Everglades wetlands have been lost to these activities. The area has experienced canal and levee building to direct the water for human use. Water that once flowed southwest now flows, in a diminished amount, to the southeast. This has resulted in the drying out of the Everglades, a decrease in the food supply, and the weakening of the system. There has been a 90% decrease in the number of wading birds in the Everglades in the last 50 years.

Because so much water has been diverted from the Everglades, the water in Florida Bay has become more saline. The salt concentration has increased to such an extent that there have been declines in the number of fish, other aquatic organisms, terrestrial animals, and plant life. The wetland nurseries are being destroyed and the whole Florida Bay system is dying. Currently, plans are underway to try to reverse some of this damage by restoring the flow of water to the Everglades. However, the removal of some of the canals and levees in an effort to save the Everglades and Florida's South Bay is being fought by agricultural and developmental interests. Once again, the whole system must be considered, not just one portion of the system. How much development and agriculture can south Florida allow when the source of its freshwater is threatened?

Irrigation for agricultural purposes places a tremendous demand on a water supply. Before water was pumped out of the ground for irrigation, farmers depended on natural precipitation. Irrigating with pumped water increases the amount of crop produced per acre and the number of acres that can be put into production. Dry-land farming is riskier and yields are lower. A farmer in Kansas needs four times as much dry-land acreage to produce the same amount of crops as could be produced on irrigated land.

Worldwide, 70% of water used for irrigation is pumped out of aquifers. Farmers are pumping water out 25 times faster than it is being replenished. These depletions are called **overdrafts**. Kansas is a natural short-grass prairie, which is now farmed using underground water from the Ogallala aquifer for irrigation. The Ogallala aquifer lies under eight high plains states, and is the largest freshwater supply of underground water in the United States. In the 1950s, using natural gas, southwest Kansas was able to tap into the Ogallala, and as a result, agriculture and cattle feed lots increased. More grain meant more cattle. All of this depended on geologic water from the aquifer. In 40 years, the water table has dropped 110 to 120 feet, and many wells have gone dry. More and more farmers are having to return to dry-land farming, which means less food production.

The use of water for irrigation has also occurred in the desert regions of the Southwest, (e.g., Arizona, southern California), where vast amounts of desert acreage are now under cultivation. But there are problems with irrigating desert soils. Under natural conditions, the amount of precipitation over a desert is low, approximately four inches per year, and this keeps the amount of vegetation sparse and the water table low.

Water contains dissolved salts. Plants take in almost pure water through their root systems, leaving the salt in the soil. The more plants you grow and the more water you use, the more salt is left on the soil. If the salt content is too high, the soil must be flushed with excess water. The extra water needed to flush the salt from the soil raises the water table. As the water table rises, water begins to evaporate from the soil surface. Only pure water evaporates, so more salts are left behind in the soil. Thus, the soil becomes even saltier, and will eventually poison the crops. This process is called **salinization**. Many agricultural desert areas are suffering soil loss by this process. A lack of understanding of how desert ecosystems work has caused this problem.

As populations grow, the demand for water for domestic and recreational use increases. These demands can come into direct conflict with other water usage demands (e.g., agriculture). The western United States is a good example of an area in which too little water is available and the demands are high.

The major source of freshwater for the seven western states is the Colorado River. The Colorado River is managed by the Hoover and Glen Canyon Dams. Water from the river is allocated to Wyoming, Utah, Colorado, New Mexico, Nevada, California, and Arizona, and Mexico. Before the dams were built, the Colorado River flowed all the way to the Gulf of California, in Mexico. Now, with the dams and the diversions of water for uses in the United States, the river dries up miles from the gulf.

In Las Vegas, Nevada, we see how the ever-increasing demand for limited water supplies will cause problems in the near future. Approximately 4,000 people per month move to Las Vegas, where an average of four inches of rain falls a year. Las Vegas is growing rapidly and needs more water. A certain amount of its water allotment comes from the Colorado River. Where will it get the water it will need in the near future? There are two possibilities. One is to siphon off more of the Colorado River, which means that someone else will have to give it up or get water from somewhere else. The second possibility is that Las Vegas might buy water from northern Nevada, which has a few small sources of underground water that could be imported through pipelines. However, those water sources sustain agriculture in the north, and with the removal of the underground water, these agricultural areas will dry up.

This approach to fulfilling a growing population's demands for water has been tried in California. Los Angeles bought underground water rights in Owens Valley, California, and piped this water to Los Angeles. Owens Valley was once one of the highest agricultural producers in California. It is now a desert. You cannot remove water from one area without having an impact on all the ecosystems of that area.

Several questions arise: Which is more important: development in Las Vegas or agriculture (food production) in the north? Are areas doing everything they can to conserve and use water wisely? Should swimming pools, man-made lakes, water parks, and golf courses be allowed in areas where water is scarce? All of these artificial structures lose thousands of gallons of water each day thorough evaporation. Should unlimited growth be allowed in areas that cannot sustain it?

Water systems also can be disrupted if the water temperature in the system is raised or lowered by only a few degrees. Aquatic organisms have narrow ranges of tolerance for temperature. When water used to cool industrial processes is dumped into an aquatic system, it is often a few degrees warmer than the water that was removed from the system. This increase in temperature can cause massive fish kills. This effect has been seen around hydroelectric plants and nuclear power plants.

Pollution degrades estuaries and coastal waters. The results of increased pollution include more frequent algal blooms, with their toxic effects; the depletion of the oxygen content of water; the clouding of the water, which blocks the sun; a decrease in photosynthesis, productivity, and diversity; and an increase in pathogens and toxins, and, thereby, an increase in disease. Sources of ocean pollution are numerous. (See Table 2-20.)

Table 2-20: Sources of Marine Pollution

Source	Share of Total
runoff and discharges from land	44%
airborne emissions from land	33%
shipping and accidental spills	12%
ocean dumping	10%
offshore mining, oil and gas drilling	1%

Source: Worldwatch Institute: State of the World, 1994.

Mining activities pollute water as heavy metals are carried into water systems. Heavy metals undergo biological accumulation and are toxic in small amounts. Marine organisms are very sensitive to chemical changes in the environment. For example, a 0.1% increase in toxic materials from sewage runoff and sludge kills herring and cod eggs.

The combination of pollution and overfishing are killing inland seas and coastal estuaries. The Aral Sea has been damaged by removing too much water for irrigation, which has made the salt content rise to pollution levels. This inland sea, which once yielded a fish catch of 44,000 tons per year, now has no commercial fish. The Caspian Sea, which 50 years ago had the largest sturgeon yield, is now only 1% of what it was because of pollution and overharvesting. The Black Sea, which once yielded 700,000 tons of fish per year, now produces only 100,000 tons, and of the 30 species of commercial fish, only five remain. The Chesapeake Bay used to yield 100,000 tons of edible oysters, but in 1993, only 1,000 tons were harvested. Thus, a decrease in the amount of freshwater slows the growth of the world food production.

What can be done to help restore the water cycle? The Clean Water Act of 1972 was a step toward cleaning up our water resources and protecting them from further damage. The law is based on the understanding that clean water is essential. It is impossible to measure the dollar benefit of clean water against the cost of cleaning it up. Therefore, everything must be done to control pollution. The law sets national water-quality standards, requires states and cities to control storm water pollution, and preserves the nation's wetlands. Unfortunately, this act is periodically revised, and it is currently under revision. Having strong protective laws is essential if ecosystems and the biosphere are to remain functional and clean water is to remain available.

Another approach to water problems is to price water at its real value. Currently, water is undervalued, therefore, it is not conserved or actively cared for. If water were environmentally priced to include the nature of the dwindling resource and the cost of clean up, people would actively conserve water and cut down on pollution.

Since clean water is required for the continuation of life, care must taken to use what water we have wisely and to keep it unpolluted. The hydrological cycle effects other nutrient cycles on a basic level, in that many elements are made available to autotrophs through their being dissolved in water. (See the phosphorous and nitrogen cycles.) Taking care of water is an investment in the continuation of life on earth.

Nitrogen Cycle

Nitrogen, as a primary constituent of proteins and nucleic acids, is an integral part of living organisms. The nitrogen cycle is a gaseous cycle that has a large reservoir pool in the atmosphere and a small active exchange pool in the soil. The rapid recycling of nitrogen is local and is closely associated with the activities of soil organisms. Nitrogen, as nitrogen gas (N_2), makes up approximately 79% of the atmosphere. Nitrogen in its gaseous form is not usable to autotrophs, and must be changed by bacteria into either ammonia (NH_3) or nitrate (NO_3^-). Nitrogen is taken in by autotrophs and passed through the grazing and detritus food chains. (See Figure 2-32.)

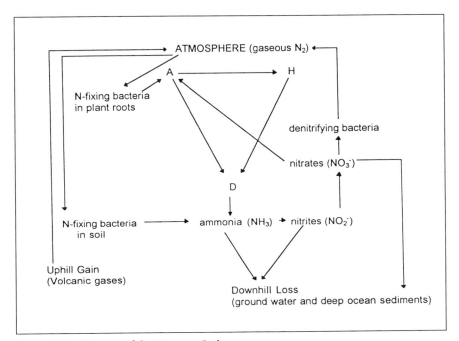

Figure 2-32: Diagram of the Nitrogen Cycle

Bacteria are essential to the recycling of nitrogen. Each chemical change is carried out by a different species of bacteria. Atmospheric nitrogen (N_2) is made available to autotrophs by **nitrogen-fixing bacteria** in one of two ways: by nitrogen-fixing bacteria associated with the root systems of certain plants and by nitrogen-fixing bacteria living free in the soil. Nitrogen taken from the atmosphere and made usable to autotrophs is said to be "fixed."

Some species of plants have evolved a relationship with certain bacteria that are associated with their root systems. These bacteria take gaseous nitrogen and convert it to usable forms. This specialized relationship occurs in the legume (pea) family and in several other families. Most gaseous nitrogen is "fixed" by free soil-living bacteria that can change it into ammonia (NH_3). The ammonia is further changed by bacteria into nitrite (NO_2^-), which is then converted to nitrate (NO_3^-). The nitrates, dissolved in the water in the soil, are then taken in by the roots of autotrophs and passed through the food chain.

Dead organic matter is broken down by decomposers, and the nitrogen is returned to the soil as ammonia, to be recycled into nitrates. Most of the nitrogen cycling through the system comes from the decomposition of organic matter. Some nitrates are converted to gaseous nitrogen (N_2) by **denitrifying** bacteria, and returned to the atmospheric pool. There is uphill gain to the system from volcanic action and downhill loss when dissolved nitrogen compounds enter deep groundwater or are carried into the oceans, to end up in deep sea sediments. Aquatic downhill losses are to deep sea sediments.

Nitrogen is often a limiting factor in terrestrial and aquatic systems, and by adding nitrogen compounds to the environment, human activity has disrupted the nitrogen cycle. The addition of excess nitrogen compounds to the environment has contributed to air pollution, acid rain, decreased autotrophic productivity, water pollution, cultural eutrophication of lakes, and global warming. The major sources of man-made nitrogen are industry waste, runoff caused by construction, fertilizer, feedlot runoff, runoff from agricultural lands, human sewage discharge, deforestation, mining, and the combustion of fossil fuels.

When fossil fuels are burned for industrial use, transport and electrical energy nitrous oxides are emitted into the atmosphere and form air pollution. (See sulfur cycle.) When nitrous oxides combine with water vapor, they form a weak acid that contributes to acid rain. Acid rain not only changes the pH of soil and water, which can pass the tolerance limits of the organisms that live there, but it can also damage photosynthetic tissue. If photosynthetic tissue is damaged, the amount of productivity for the system decreases. (See tolerance limits.) If soil becomes too acidic, the bacteria essential for recycling nitrogen can be killed. Nitrous oxides are an important greenhouse gases responsible for global warming. (See carbon cycle.)

Any human activity that results in increased soil erosion increases the amount of water runoff from a system. It also results in an increased amount of nitrogen compounds, which are dissolved in the water, being carried away. As nitrogen leaves the system, the system's productivity decreases.

Agriculture has been a primary cause of the removal of nitrogen from terrestrial systems. Crops take in nitrogen and convert it to plant biomass. When the crops are harvested, the nutrients that they have incorporated leave the system. With continued agricultural use, the soil becomes nitrogen poor and, to compensate for this, fertilizer is added. Fertilizers are used as an energy subsidy to increase crop production. Unfortunately, a large amount of the nitrogen in fertilizer ends up in the water system and leaves the area in the ground water.

Nitrogen, along with phosphorous, pollutes the river and lake systems that it enters. The water systems that receive this additional nitrogen and phosphorous are enriched. This enrichment is called **eutrophication**. Eutrophication caused by human activity is called cultural eutrophication. The primary sources of **cultural eutrophication** are industrial waste, fertilizers, and human sewage. The excess nitrogen acts like a fertilizer to the aquatic system, and can kill the system.

How does this happen? With more nitrogen in the system, the number of algae increases. As this population increases, the demand for oxygen for cellular respiration rises. When the algae die, they fall to the bottom of the lake, with the result that the decomposer population increases because of an increased food supply. More decomposers demand more oxygen for cellular respiration. Since there is less dissolved oxygen at the bottom of the water column than there is at the top, all of the oxygen is used up. When this happens, the entire lake system is disrupted and the lake dies.

Several things can be done to return the nitrogen cycle to normal. The amount of nitrogen that is released into the atmosphere can be reduced by decreasing the amount of fossil fuel burned. Since much fossil-fuel burning is directly linked to energy production, the conservation of fuels and the use of alternative fuels can decrease nitrous oxide production. The amount of nitrogen that ends up in aquatic systems can be decreased by decreasing deforestation, changing dietary patterns (less beef = fewer feedlots = less runoff of waste), and decreasing soil erosion. Nitrogen-fixing plants can be used to enrich agricultural soils, thereby decreasing the need for commercial fertilizers.

Better control of human sewage treatment and discharge are also needed. Because nitrogen is a limiting factor in most systems, adding more nitrogen to the system would increase productivity. However, it is not always better to have more nitrogen, unless you can add more without disrupting the system.

Phosphorous Cycle

Phosphorous is part of the structure of nucleic acids, phospholipids, adenosine triphosphate (ATP), bones, and teeth. Phospholipids are important because they form the basic structure of cell membranes, and ATP is the energy currency used to run the cellular chemistry. The phosphorous cycle is a sedimentary cycle with a large reservoir pool in the earth's crust and a small active exchange pool in the soil. Phosphorous is scarce on the earth's surface, and there is no atmospheric component to the cycle. Phosphorous is cycled locally by the biotic interactions of the grazing and detritus food chains. The element is slowly released from the reservoir pool by the weathering and erosion of rocks and is relatively unavailable. Therefore, the pool is less self-adjusting than a gaseous cycle and can be easily disrupted. (See Figure 2-33.)

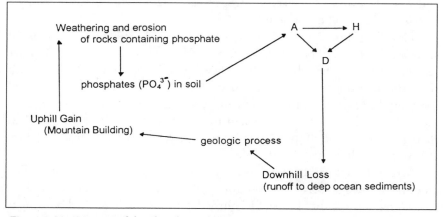

Figure 2-33: Diagram of the Phosphorous Cycle

Phosphorous is taken up from the soil solution by the roots of autotrophs as orthophosphate ($H_2PO_4^-$). It is then incorporated into autotrophic biomass and passed through the food chains. The decomposers return the phosphate to the soil by breaking down organic matter. Because phosphates are dissolved in water, anything that causes increased runoff will affect the cycle by removing the phosphate from the local system. The phosphates carried away in runoff eventually end up in the ocean, and, with other elements such as calcium and sodium, settle into the deep sea sediments. This downhill loss of phosphorus is slowly reversed when mountain building occurs, lifting up deep sea sediments.

Phosphorous is often a limiting factor in terrestrial and aquatic systems, and by adding phosphorous compounds to the environment, human activity has disrupted the cycle. The addition of excess phosphorous compounds is usually in the form of fertilizer, detergents containing phosphorous, and human and animal waste. The increase of phosphorous contributes to water pollution and the cultural eutrophication of lakes. Fertilizer runoff is the primary source of aquatic contamination. As world food demands have increased, so has the use of fertilizers. (See Figure 2-34.)

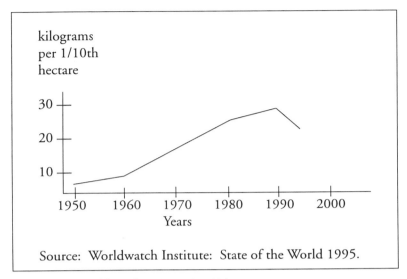

Figure 2-34: World Fertilizer Use 1950-1994

There is a physiological limit to how much fertilizer a crop can use. The dip in world fertilizer use and the fact that many countries have reached their limit have been reflected in a more stable rate of food production The phosphate rock used for fertilizer production is not uniformly distributed in the biosphere: 73% of the world's phosphate rock is mined in the United States, the Soviet Union, and the Moroccan Sahara.

Additional phosphorous in a system has the same effect as nitrogen on aquatic systems. The same human activities that increase nitrogen runoff increase the removal of phosphorous as it is dissolved in water. (See nitrogen cycle.)

The pH of the soil can affect the availability of phosphorous. Below a pH of 5.5, phosphorous reacts with iron and aluminum, forming insoluble compounds. Above a pH of 7.0, it forms insoluble complexes with calcium. Acid rain can affect the availability of phosphorous as it changes the soil's pH.

Several things can be done to return the phosphorous cycle to normal. The amount of phosphorous that is released into aquatic systems can be decreased by changing dietary patterns (less beef = fewer feedlots), the amount of deforestation can be reduced, and the treatment of human sewage treatment and discharge can be controlled. The use of agricultural technologies such as limited till and crop rotation to enrich soils, instead of the application of commercial fertilizers, would decrease the amount of phosphates in the runoff. Though phosphorous is a limiting factor in terrestrial and aquatic systems, more is not always better.

Sulfur Cycle

Sulfur is a building component of proteins, which are essential to living organisms. The sulfur cycle is a sedimentary cycle with a large sedimentary reservoir and a small atmospheric reservoir. Sulfur moves rapidly through the food chains of the local ecosystem by biotic activity. In the atmosphere, sulfur is in the form of sulfur dioxide (SO_2); it is used by most autotrophs in the form of sulfate (SO_4^{-2}). Autotrophs take sulfate in through their root systems, incorporate it into the plant biomass, and pass it through the food chains. When organic material is broken down by decomposers, its sulfur is released into the soil as hydrogen sulfide (H_2S). Hydrogen sulfide is then converted by bacterial action into sulfate or sulfur dioxide. Some sulfur is returned to the soil as sulfate in heterotrophic waste. There is an uphill gain to the atmospheric pool from volcanoes and hot springs, and a downhill loss to deep sea sediments. (See Figure 2-35.) Sulfur is seldom a limiting factor in terrestrial or aquatic systems.

Human activity has increased the amount of sulfur in the atmosphere. This sulfur is in the form of sulfur oxides (SO_x) and is a major source of **air pollution**. **Air pollution** is the unfavorable changing of the atmosphere by human activity. Sulfur and nitrous oxides form one-third of the industrial pollutants in the United States. Sulfur oxides are produced primarily by the burning of coal by large power plants; nitrous oxides are primarily emitted from car exhausts when fossil fuels are combusted. When these oxides combine with water they form weak acids (sulfuric and nitric acids) that contribute to the formation of **acid rain**. (See tolerance limits.) Acid rain changes the pH of terrestrial and aquatic ecosystems, stressing their biotic component and affecting the systems' ability to function normally. When ultraviolet radiation interacts with NO_x and unburned hydrocarbons, photochemical smog is produced.

Acid rain and **photochemical smog** contribute to decreased autotrophic productivity (crop production) by damaging delicate photosynthetic tissue. Air quality also directly affects the health of heterotrophs. Pollutants cause respiratory problems because air exchange tissues are delicate tissues and are easily damaged. The droplets of sulfuric and nitric acid are drawn deep into

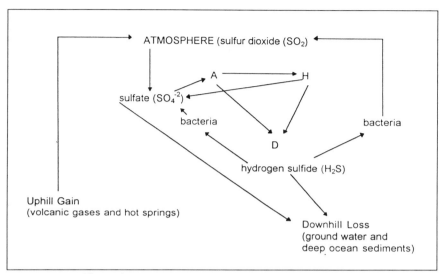

Figure 2-35: Sulfur Cycle

the lungs as an acid mist. The number and severity of respiratory diseases (e.g., asthma, chronic bronchitis, emphysema, lung cancer), increases with the decrease in air quality and the length of exposure.

The **Clean Air Acts** of 1963 and 1970 addressed the issue of local poor air quality. The response of the industries to the Clean Air Acts was to build higher smokestacks to carry pollutants away from locally affected areas. Taller smokestacks were quick, cheap, local fixes to the problem. In 1956, the average height for smokestacks was 60 meters. By 1976, the average height ranged from 250 to 360 meters. Air quality in the local areas did improve because the pollution was carried downwind. Higher smokestacks did not decrease air pollution, they moved it. What is needed is a long-term fix.

How can the amount of air pollution be reduced? How can air quality be improved? By not producing air pollutants and/or not emitting them into the atmosphere. Industries must clean up what they emit, as well as find new ways to manufacture products that do not create pollution in the first place. Since the burning of fossil fuel is a primary source of air pollutants, alternative fuel sources for electricity and for vehicles need to be developed and used. Until these are available, a simple decrease in energy use would decrease the amount of fossil fuel burned.

Air pollution problems are often discussed in terms of individual pollutants and their effects. However, the effect of pollution on an organism or a system is not just the sum of the effects of each pollutant. The total effect is greater than the sum of the damage done by separate pollutants. The effect of two or more pollutants acting together is called **synergism**.

Oxygen Cycle

Oxygen is a basic requirement for life on earth today. Although there are a few bacteria that do not use oxygen, all other autotrophs and heterotrophs require a steady supply. Without it, they die. Oxygen is required for autotrophs to carry on photosynthesis. Autotrophs and heterotrophs require oxygen for the process of cellular respiration. Cellular respiration breaks down carbon compounds to release the energy that drives cellular activities. The oxygen cycle is a gaseous cycle with a large atmospheric reservoir. The atmosphere is 21% gaseous oxygen (O_2). (See Figure 2-36.)

There are uphill gains to the cycle from volcanic activity and downhill losses to deep sea sediments. When excess carbon is removed from the carbon cycle by the Atlantic Engine (See carbon cycle.), it is combined with oxygen. The loss of carbon to deep sea sediments, in the form of calcium carbonates, means the loss of oxygen. The cycles are directly connected.

Atmospheric oxygen is also important to living systems because of the part that it plays in the formation of the **ozone layer** in the lower stratosphere. **Ozone** (O_3) is formed naturally when molecules of gaseous oxygen (O_2) interact with solar radiation. The ozone layer absorbs most of the ultraviolet radiation (UV) from the sun and acts as a shield for the biosphere. UV radiation is high-energy radiation that can break the hydrogen bonds that hold biological molecules together. The organisms that live on earth today evolved in an environment in which a high level of UV radiation was not a selective pressure. Therefore, they are not adapted to high levels of UV radiation.

Most organisms that live on the earth's surface have some form of protection for their protoplasm (e.g., hair, feathers, scales, shells). High levels of UV radiation can kill unprotected protoplasm and damage protected protoplasm. Earthworms, which normally live in the soil, have no extra protection against UV radiation. Often after a summer rain, you will see dead earthworms on the sidewalk. This occurs because the rain has saturated the soil, filling all the air spaces. As a result, the worms have come to the surface because they need oxygen, the sun has come out, and the worms have been killed by UV radiation.

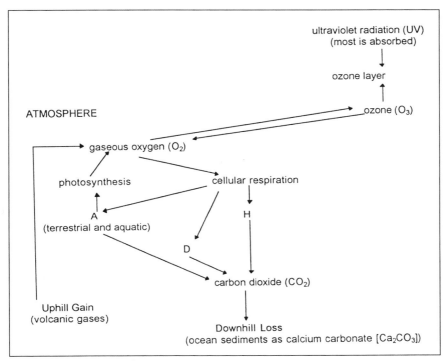

Figure 2-36: Oxygen Cycle

Human activity (industry) has produced products that have released materials into the atmosphere, causing the destruction of the ozone layer. This depletion, or thinning, is called the "**ozone hole.**" There is a natural thinning of the ozone layer annually over the Arctic and the Antarctic, but this natural thinning has been accelerated by human activity. Some chemicals release chlorine when they break down in the atmosphere. The chlorine interacts with the ozone and produces gaseous oxygen, thereby destroying the ozone layer. Once released into the atmosphere, chlorine can remain and do damage to the ozone layer for 40 to 100 years. (See Table 2-21 for a partial list of ozone destroyers.) The amount of chlorine produced by the U.S. chemical industry increased 100-fold between 1920 and 1990. The U.S. produces 10 million tons of chlorine a year, one-third of the world's total chlorine production.

Table 2-21: A Partial List of Pollutants That Destroy the Ozone Layer

chlorofluorocarbons (CFCs)
hydrochlorofluorocarbons (HFCs)
carbontetrachloride
methychloroform

Increased UV radiation has serious health implications, from causing skin cancer and cataracts to suppressing the immune responses of organisms. As the ozone level decreases, the size of the normal ozone hole increases (the amount of thinning increases). The effects of this thinning are already seen in Australia and South America, where the incidences of skin cancer and cataracts are on the rise. A 1% decrease in ozone can produce a 2- to 5% increase in the number of human skin cancers. As thinning of the ozone layer increases, it will affect the ocean's productivity. As the amount of UV radiation increases, phytoplankton, which live near the surface of the water, go deeper to get away from the UV. This means they receive less radiant energy, and the amount of photosynthesis and productivity decreases. This decrease in productivity affects the food web of the system.

Ozone also can be a problem when it forms in the lower atmosphere or at ground level. Low-level ozone is a component of air pollution, causing respiratory disease and damage to photosynthetic tissues, thereby decreasing photosynthesis and productivity. Too much ozone at ground level acts as an air pollutant, and too little ozone in the ozone layer causes increased UV radiation problems.

What can be done to correct these ozone problems? The ozone layer will slowly reestablish itself if chlorine is no longer emitted into the atmosphere. Therefore, chlorine compounds must be eliminated. Industry must find new ways to manufacture and new materials that do not damage the environment. Many countries have curtailed the use of CFCs by no longer using them in aerosol sprays and refrigerants. But many other chlorine products are still being used and continue to affect the ozone layer. We cannot make or rebuild the ozone layer; we can only stop destroying it and let it reform naturally.

Toxic Material

Toxic materials, or toxic substances, are chemicals that can adversely affect individuals and systems. These materials can be natural or man-made. (See Table 2-22 for a partial list of toxic materials.) The naturally occurring toxic materials include heavy metals, such as mercury and lead. Under most natural conditions, these metals are in a sufficiently low concentration that they do not affect organisms. But human activity has increased the concentration of these materials to dangerous levels. Many undergo biological magnification once they are in a system. Industrial waste and mining runoff contain high levels of heavy metals.

Table 2-22: A Partial List of Toxic Materials

asbestos
heavy metals: lead, tin, mercury, cadmium
polychlorinated biphyls (PCBs)
pesticides
herbicides
fungicides
solvents

There are over 50,000 industrial chemicals made, used, and emitted into the environment in the United States. Toxicity information is only available for about 20%, and this information is incomplete.

Toxic materials affect the central nervous systems and the immune systems, and cause cancer and respiratory problems when they enter the lungs with other air pollutants. The individual chemical hazards and their synergistic effects are little understood. The long-term effects of low-level exposure to toxic materials are not known, and the combined effects are undetermined. These materials may be emitted into the environment locally, but they do not stay there. (See the water cycle.) Once again, the systems must be viewed as a whole, not as a group of individual parts. The precautionary principle has not been adequately applied to the chemical industry and its products.

Nuclear waste comes under the heading of toxic material. Nuclear waste is primarily the product of nuclear weapons testing and use, and of the nuclear power industry. Nuclear waste gives off high levels of high-energy radiation that is damaging and/or lethal to living protoplasm. Radiation exposure affects adult organisms, as well as the next generation. Radiation exposure causes mutations in sperm and eggs and results in birth defects. Adult organisms exposed to a higher than normal dosage of radiation show increased rates of cancer. Nuclear waste is a unique problem because it emits dangerous levels of radiation for long periods of time. The half-life of spent nuclear plant fuel rods is 7,000 years. At present, there is no safe long-term way to store or dispose of nuclear waste.

Recycling

What happens to all of the materials that you and I throw away? Until recently, all of the garbage produced (solid waste) ended up in landfills or in the oceans, or it was incinerated. All of this solid waste is composed of materials made from natural resources (raw materials) by either nature or man. Natural resources are any substances used by man that are derived from the environment. **Natural resources** are divided into three categories: **perpetual**, **renewable**, and **nonrenewable**.

Perpetual resources are found in the environment in such large quantities that they are essentially inexhaustible (e.g., sand, seawater, sunlight). **Renewable resources** are produced at a rate greater than or equal to the rate of consumption (e.g., most foods, atmospheric oxygen). **Nonrenewable resources** are produced at a rate less than the rate of consumption (e.g., fossil fuels, iron, aluminum, copper, gold and most other metals, groundwater). When we think about resources, we need to consider how much of each resource we are using, how fast we are using it, and how much is left. **Reserves** refers to the total amount of a resource that can be used in a cost-effective manner.

The needs to preserve natural resources, reduce waste, conserve energy, and decrease toxic waste all contribute to the increased push toward recycling and reuse. Between 1970 and 1991, the total world consumption of materials increased by 38%, forestry 44%, metals 26%, nonmetallic minerals 39%, and nonrenewable organic chemicals 69%. Because many materials are being used faster than they can be produced, and because we are running out of places to put all of our waste, the practices of recycling and reusing materials have become necessary. Landfills are reaching capacity, oceans are being polluted and showing signs of malfunctioning, and incinerators continue to produce and emit toxic pollution.

Many man-made products are not **biodegradable**. These materials are not broken into natural materials by decomposers. A glass bottle will not break down for 300 years. *Biodegradable* does not mean that a product fragments into smaller pieces of itself. Some "biodegradable" plastic garbage bags do fragment into smaller pieces of plastic, but they are not broken down into natural materials by decomposers. **Recycling** means "to put materials back into use, either by reclaiming the material and using it again or by reusing the material in the same or a different form."

Raw materials industries are one of the world's biggest consumers of energy. Paper, steel, aluminum, plastics, and container glass manufacturers use 37% of the energy that is used annually in the United States. The four primary production industries (paper, plastics, chemical, and metal) produce 71% of the toxic emissions in the United States. The use of raw materials, (extraction, processing, refining, manufacturing) all have environmental costs. By recycling, reusing, and reducing consumption, these costs can be decreased. For example, if you were to recycle a stack of newspaper 36 inches high, you would save one tree, decrease by 30- to 55% the amount of energy needed to produce new paper, and decrease by 95% the air pollution from pulp mills. (See Table 2-23 for the amount of energy used in the production and recycling of some building materials.)

Table 2-23: Energy Used in the Production and Recycling of Selected Building Materials in the United Kingdom

Material	Virgin Production (gigajoules per ton)	Recycling (gigajoules per ton)
Brick	2.5-6.1	not recycled
Wood	4-5	not recycled
Glass	13-25	10-20
Plastics	80-220	50-160
Steel	25-45	9-15
Copper	70-170	10-80
Aluminum	150-220	10-15

Source: Worldwatch Institutes: State of the World, 1995.

   Solid waste is the term used for the waste material that is regularly collected from households, institutions, agriculture, industry, and commercial establishments. The United States contains only 10% of the world's population, but it produces one-third of its waste. This figure indicates our overconsumption of materials and lack of conservation.

   Household garbage content can be broken down into the following categories: food waste, glass, yard waste, metal, plastics, paper, and other. The following are a few examples of the waste materials in the United States and the savings possible from recycling.

1. *Paper:* The United States uses approximately 67,000 tons of paper each year, or about 600 pounds per person. It takes 17 trees to make one ton of virgin paper: about 2.3 of an acre of forest. By recycling one ton of paper, you can save 17 trees, three cubic feet of landfill space, two barrels of oil, 7,000 gallons of water, and 4,100 kilowatts of electricity (enough to power the average home for five months). Recycling of paper saves 62% of the energy required to make paper out of trees.

2. *Glass:* Approximately 41 billion glass containers are produced in the United States each year. Glass is 100% recyclable. It can be collected, remelted, and molded into new containers. Recycling one glass bottle saves enough energy to light a 100-watt bulb for four hours. Each ton of recycled glass saves eight gallons of fuel oil.

3. *Lead-acid Batteries:* Although these batteries make up only a small percent of the waste stream locally, nationally they equal about 40,000 tons. Over 78 million lead-acid batteries are thrown away annually. Both the acid and the lead from these batteries can be recycled, reducing toxic waste runoff into aquatic and terrestrial systems.

4. *Household Batteries:* These batteries contain heavy metals (e.g., mercury, cadmium, lead, lithium, manganese, silver, nickel, zinc), all of which are toxic in high concentrations. Each U.S. household uses about 32 batteries annually. About 83,500 tons of these batteries enter the waste stream each year. They are a source of aquatic and terrestrial pollution. Recycling decreases pollution and the costs to the environment from the extraction of raw materials.

5. *Motor Oil:* Used motor oil is the single largest source of oil pollution in the waterways of the United States. About 200 million gallons of used motor oil end up each year in our soil and water. About one million gallons of used motor oil are dumped into the environment each day. A

total of 1.3 million barrels of oil would be saved annually if all used motor oil were recovered and recycled. Rerefining oil takes one-third the amount of energy that the refining of crude oil takes, saving energy and fossil fuel. Recycling motor oil decreases the aquatic and terrestrial environmental damage due to dumping.

6. *Aluminum:* The average American generates five pounds of aluminum waste each year. Aluminum is 100% recyclable. One recycled can saves enough energy to burn a 100-watt bulb for 3-1/2 hours. Recycling aluminum saves 95% of the energy necessary to obtain aluminum from raw materials. Recycling saves energy, raw materials, and fossil fuel.

7. *Plastics:* Plastic containers, packaging, and wrap account for about eight percent of the volume of solid waste in landfills. Recycling plastics saves on fuels, materials, and landfill space. These materials are not biodegradable.

8. *Steel:* Steel is another product that is 100% recyclable. For each ton of steel recycled, 2,500 pounds of iron ore, 1,000 pounds of coal, and 40 pounds of limestone are saved. Recycling steel cans helps to extend the life of the landfills, saves energy and raw materials, and decreases pollution.

9. *Tires:* In the United States, about 240 million used tires are discarded annually. These tires can be used an alternative energy source, in that they produce more energy than fossil fuel when they are burned. They also produce less ash and sulfur emission (pollution) than coal.

The following are several suggestions that could increase the practices of recycling and reusing materials, reduce the waste stream, and decrease consumption.

1. Raw materials should be taxed at their full environmental cost, not subsidized. The lumber and mining companies that harvest raw materials from public lands should pay their full environmental cost, not extract them at reduced cost (subsidies).

2. Everyone should pay the full cost of disposing of wastes. This practice would encourage recycling and decrease consumption of goods.

3. Redundant packaging of products should be eliminated. This would decrease the use of raw materials and decrease solid waste.

4. Waste products should be transformed into new products. This would decrease landfill and raw materials usage.

5. Food and yard waste should be composted on-site. This would not only decrease the solid waste stream, but the compost produced could also enrich the soil.

6. The consumerism and the throw-away philosophies of society should be altered. These changes in approach to product demand and usage would greatly reduce the stress on the biosphere in almost every area of concern from water to air pollution.

7. Products should be produced for quality, not quantity. Product lifetimes need to be increased so that goods last longer, putting a stop to built-in obsolescence.

8. The green pricing of products should be increased so consumers can make informed decisions.

9. The practice of sustainable yield should be encouraged. **Sustainable yield** means that a substance is harvested, or extracted, at a rate equal to the yearly production rate of that substance in the system. For example, the amount of lumber cut in an area each year should equal the amount of wood that the forest can produce each year.

10. The use of rechargeable batteries should be encouraged.

11. The manufacture and use of products that help the environment should be encouraged. Compact fluorescent lights last longer than incandescent lights, and are more energy efficient. This means that the waste stream is decreased and energy is saved. Saved energy means less fossil fuel is burned and less air pollution is produced.

ASSIGNMENTS FOR PART 2: ECOSYSTEM
ABIOTIC COMPONENT

Limiting Factors

1. Explain to your friend why you find polar bears in Alaska and not in California.

2. Explain how the Law of Tolerance relates to the Theory of Evolution.

3. Refer to Figure 2-11 and use a specific example to explain why there are two zones of physiological stress.

4. Why is it important to understand the complete life cycle of an organism when discussing its limits of tolerance?

5. What is acid rain and how does it relate to the Limiting Factors Principle?
6. Explain to a friend what ecological indicators are and how they are used by ecologists.

Energy

1. Explain to a friend the first and second laws of thermodynamics.
2. Explain how the second law of thermodynamics relates to entropy.
3. Explain why the second law of thermodynamics is important in the structure and function of an ecosystem.
4. Choose a specific example and explain the importance of thermal flux.
5. Write an essay comparing photosynthesis and cellular respiration.
6. Explain to a friend how primary productivity differs from secondary productivity.
7. Why is net primary productivity important to the functioning of an ecosystem?
8. Under which environmental selective pressures would C_4 plants have evolved?
9. Explain why some countries require more ghost acreage than others.
10. Explain to a friend what the green revolution was.
11. Why doesn't the green revolution hold the answer to feeding the expanding world population?
12. Explain why using the tropical rainforests for agricultural purposes, aquaculture, and hydroponics will not be the solutions to the world food problem.
13. Explain to a friend the differences between a grazing food chain and a detritus food chain.
14. Construct a chart comparing the ecological pyramids for a grassland and a forest.
15. Using a specific example of your choice (not DDT) explain how biological magnification works.
16. Use Figure 2-15 to explain to a friend the United States corn yield between 1866 and 1993.
17. Construct a graph of the information in Table 2-8 on world grain production and fertilizer use between 1950 and 1993. Explain how this relates to the world food problems.
18. Construct a graph of the information in Table 2-9 on per capita grain use and consumption of livestock products in 1990. What conclusions can you draw form your graph?

19. What conclusions can you draw from Figure 2-17 on energy subsidies for food production?
20. What conclusions can you draw from Figure 2-23 on world grain use, total, and for feed between 1960 and 1992?
21. Construct a graph of the information in Table 2-12 on DDT.

Soil

1. Explain to a friend how soil is formed.
2. Use a specific example of your choice and explain why permeability and capillary water are important to plants.
3. Go into your backyard, or another area that allows digging, dig a hole (soil pit), and draw the soil profile. Describe the texture, feel, appearance, moisture content, and depth of each horizon. What kinds of things did you find in the litter on top of the soil?
4. Explain why the chemical composition of the soil is important to the ecosystem.
5. Explain to a friend why changing the pH of the soil can damage the structure and function of the ecosystem.
6. If you were to use a pesticide on your lawn, what effect could it have on the soil's biotic community and why would this be important?
7. Compare the three classifications of soil according to age and productivity.
8. Explain how the use of the tropical rainforests for food production is a good example of the precautionary principle.
9. What are the two natural causes of soil erosion and how did they contribute to the Dust Bowl of 1934?
10. Discuss the agricultural changes that were instituted after the Dust Bowl to improve the soils.
11. Discuss how human activities have contributed to the current loss of soils and decreased soil fertility in the United States.
12. Using an example of your choice, explain to a friend the advantages and disadvantages of having a taproot system versus a fibrous root system.
13. Using the information in Table 2-13 on worldwide land degradation, determine for each region which activity causes the most land degradation.
14. Using the information in Table 2-13, construct a graph of worldwide land degradation due to human activity. What conclusions and inferences can you draw from your graph?

15. Using the information in Table 2-13, compare the contribution of various human activities to the land degradation in North and Central America with that of Asia and Africa.
16. Construct a graph using the information in Table 2-13. Graph the percentage that each human activity contributes to the land degradation for Europe and North and Central America. What conclusions and inferences can you draw from your graph?

Biogeochemical Cycles

1. Compare the two pools found in each nutrient cycle.
2. Compare sedimentary and gaseous cycles.
3. Explain to a friend the difference between turnover rate and turnover time. Why is one more important than the other in understanding the function and structure of an ecosystem?
4. Using an example of your choice, explain global circulation and why it is important.

Carbon Cycle

1. Draw a pictorial illustration of the carbon cycle.
2. What causes an ice-age to begin every 200,000 years?
3. Explain to a friend the effect that deforestation has on the carbon cycle.
4. Explain how clear cutting is different from selective cutting. How do they effect the ecosystem and the biosphere?
5. Construct a graph to show tree size in diameter for the harvesting of an old-growth forest and for five harvesting cycles on a tree plantation. Explain why your graph looks the way that it does. [Note: The harvesting time on tree plantations is 60 years.]
6. What inferences and conclusions can you draw from Figure 2-28 on world paper consumption from 1931 to 1991?
7. What is environmental pricing and why is it important? How would it change your life?
8. What inferences and conclusions can you draw form the Figure 2-29 on U.S. materials consumption and populations from 1900 to 1991?
9. Construct a graph of the information in Table 2-15 on the pounds of CO_2 produced based on the types of fuel.

10. Using the energy bills and gasoline consumption for your family, construct a graph of the amount of CO_2 produced for a two-month period. What could you change to decrease your CO_2 production? Try these strategies for a month and compare your new consumption with your old patterns. What conclusions can you draw?
11. Explain to your friend the differences between the greenhouse effect and global warming.
12. Graph the information found in Table 2-16 on important greenhouse gases. What conclusions and inferences can you make from your graph?
13. Describe to a friend how the Great Conveyor and Atlantic Engine work to reduce global CO_2.
14. What will be the effect of global warming on the biosphere?
15. How will global warming affect you?
16. What changes can be made to decrease the amount of greenhouse gases in the atmosphere?

Water Cycle

1. Draw a pictorial representation of the water cycle.
2. Explain the importance of the information in Table 2-17.
3. What is the different between ground water and surface water?
4. Construct of graph of the information in Table 2-18. What conclusions and inferences can you make from your graph?
5. Using the information in Table 2-19, calculate how much water you and your family, or your roommate, use each day. Plan how you can decrease the amount of water you use and implement your plan. Calculate your new water usage and compare the results with your original usage.
6. Using specific examples, explain to a friend how human activities have affected the water cycle.
7. Explain to a friend the importance of maintaining wetland areas.
8. What is an aquifer and why is it important to understand the effect of overdrafts?
9. What is the process of salinization and how does it effect world food production?
10. Explain the importance of the Clean Water Act of 1972 to the hydrological cycle. Why was it necessary for Congress to pass this act? Why is it important for this act to stay in effect? Find two current articles in the library that deal with the importance of clean water and summarize them.

11. How do forests help to maintain the water cycle?
12. Using specific examples, explain to a friend why communities need well-formulated development plans.

Nitrogen Cycle

1. Draw a pictorial illustration of the nitrogen cycle.
2. Explain to a friend the importance of bacteria to the functioning of the nitrogen cycle.
3. Summarize how human activities effect the nitrogen cycle.
4. Explain to a friend what cultural eutrophication is, what causes it, and how it affects aquatic systems.
5. What do you do in your life to disrupt the nitrogen cycle? What can you do to decrease your impact on the cycle?
6. How does the holistic approach relate to the nitrogen cycle? (Refer to Part 2: System to review the meaning of *holistic.*)

Phosphorous Cycle

1. Draw a pictorial illustration of the phosphorous cycle.
2. Compare either the carbon or nitrogen cycle with the phosphorous cycle.
3. What inferences and conclusions can you draw from Figure 2-34 on world fertilizer use from 1950 to 1994?
4. Explain to a friend how acid rain can affect the phosphorous cycle.
5. What do you do in your life to disrupt the phosphorus cycle? What can you do to decrease your impact on this cycle?
6. How does the precautionary principle relate to the phosphorous cycle?

Sulfur Cycle

1. Draw a pictorial illustration of the sulfur cycle.
2. Go to the library and find two articles on the effect of air pollution. Summarize what you learned in the articles.
3. Find two articles in the library on the Clean Air Acts and summarize the information they present.
4. Explain to a friend how human activity has affected the sulfur cycle.
5. Explain to a friend what air pollution is and how it affects autotrophs and heterotrophs.

Oxygen Cycle

1. Draw a pictorial illustration of the oxygen cycle.
2. Explain to a friend the importance of the ozone layer, how it is formed, and how human activity is affecting it.
3. Find two articles in the library that discuss the ozone depletion problem and summarize them.
4. How does ground-level ozone affect individuals and ecosystems?
5. Find two articles in the library that discuss the causes of ground-level ozone. Summarize what you learned in the articles.

Toxic Materials

1. Explain to a friend the affect of toxic materials on individual organisms.
2. Find two articles in the library that deal with the effects of toxic materials on the biosphere. Summarize the information you learned in the articles.
3. Explain to a friend why nuclear waste is a problem now and for future generations.
4. Explain to a friend why the precautionary principle should be applied to the chemical and nuclear industries.
5. Make a list of the toxic materials you are exposed to in your home, school, and work environments.

Recycling

1. Contact your local municipality and find out how much waste is recycled in your town, and how the program works.
2. Contact your county and ask for a complete set of recycling fact sheets and for information on hazardous waste recycling.
3. Keep track of how much and what types of solid wastes you produce in a week. Make a plan to decrease your waste production, carry out the plan, and calculate the amount of waste reduction you achieved.
4. Explain to a friend why recycling and reuse are important.
5. Explain to a friend how recycling affects the environment.
6. Graph the information in Table 2-23. What conclusions and inferences can you draw from your graph?
7. List five examples of each type of natural resource (perpetual, renewable, nonrenewable) that you use daily.

SUGGESTED READINGS

Carson, Rachel. (1961, 1991). *The sea around us.* New York: Oxford University Press. (warning of environmental ocean problems)

Carson, Rachel. (1962). *Silent spring.* Boston: Houghton Mifflin. (a warning of pesticide impact on the environment)

Krebs, Charles. (1994). *Ecology* (4th ed.). New York: Harper-Collins. (ecology textbook)

Lopez, Barry. (1986). *Arctic dreams.* New York: Bantam Books. (naturalist essays about the arctic environment, people, plants, and animals)

Odum, Eugene. (1971). *Fundamentals of ecology* (3rd ed.). New York: W.B. Saunders, Harcourt Brace & Co. (ecology textbook)

HOW DOES THE ECOSYSTEM CHANGE?

Communities change over time in a predictable sequence of stages, or phases. This change is called **ecological succession, community development**, or **ecosystem development**. Each stage in this series of changes is called a **seral stage**.

Why do systems change over time? The first species to move into an area are called **pioneer species**. They modify the physical environment by increasing the water retention and nutrient content of the soil and the amount of shade available, and by changing the temperature of the soil. As the physical environment changes, the biotic community changes. In a newly abandoned corn field, the first species to move into the field (colonize it) can tolerate a lot of sun, poor soil with low nutrient levels, and a wide range of temperature fluctuation. Over time, these species change the field's physical conditions so much that new species can move in and out-compete the first species. The new species have different tolerance limits and are better adapted to the new physical conditions. Each species prepares the way for the next species, actually doing itself out of its home.

This process continues and the system continues to change. These generalist colonizing species are opportunist (*r*-selected) species with wide ranges of tolerance. The final species in the mature system are specialists: equilibrium (*K*-selected) species with narrow ranges of tolerance. The beginning seral stages are the pioneer stages and the final seral stage is the **climax stage**. The climax stage is not replaced by any other community. It is self-perpetuating and unchanging in its species composition. The composition of the climax community depends on temperature, altitude, amount of radiant energy, and seasonal changes in temperature and patterns of precipitation. (See Figure 2-37.) The climax stage for northern New Jersey is the Eastern deciduous forest, which is dominated by oak, hickory, and maple species. The climax stage in Iowa is the prairie, which is dominated by grass species. Table 2-24 shows some of the changes in community characteristics that occur during succession.

There are two types of succession: **primary** and **secondary. Primary succession** occurs on sterile sites where no previous community has existed (e.g., rock outcroppings of parent material, cooled lava flows, areas scraped clean by retreating glaciers). There is no soil and no living organisms are present.

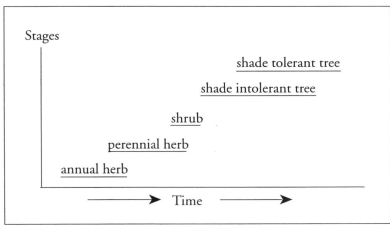

Figure 2-37: Stages of Terrestrial Succession

Table 2-24: Changes in Community Characteristics During Succession

Characteristic	Change	
	Pioneer Stage	Climax Stage
Productivity	High	Low
Respiration Cost	Low	High
Net Productivity	High	Low
Biodiversity	Low	High
Niches	Few	Many
Food Web	Simple	Complex
Community Stability	Low	High
Biomass	Small	High
Number of Species	Small	Large
Tolerance Limits	Broad	Narrow
Life Cycle	Simple	Complex
Length of Life	Short	Long
Organization	Simple	Complex
Energy Efficiency	Poor	Good
Nutrient Utilization	Poor	Good

The first pioneer species are lichens, which are composed of fungi and algae. The fungi protect the algae from drying out and the algae carry on photosynthesis, making organic compounds useful to the fungi. Over time, the lichens die and decompose, the parent material weathers, and soil begins to accumulate. When enough soil has accumulated, other pioneer species move into the site and succession continues. Primary succession is very slow and the rate of succession depends on the rate of soil development. (See Figure 2-38.)

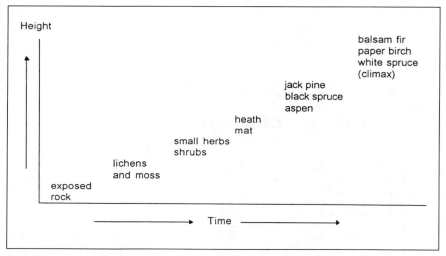

Figure 2-38: Generalized Diagram of Primary Succession after Several Hundred Years in Northern Lake Superior

Secondary succession occurs on sites where a previous community has been established. The site has suffered some form of disturbance or natural disaster that has removed the established community (e.g., fire, flood, bulldozer, cultivator). Although the community has been removed, the soil is intact. Secondary succession proceeds more quickly than primary succession because soil is present and some seeds, spores, roots, and soil organisms are already on-site. The first species to move into the disturbed site will be generalists, followed by specialists. (See Figure 2-39.)

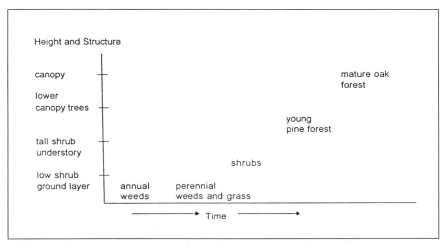

Figure 2-39: Generalized Diagram of Secondary Succession in the Northeast over Approximately 150 Years

The order of change through the seral stages is predictable, but the composition and length of each stage will depend on the original community and the physical conditions. Some early seral stages may not occur. Different ecosystems reach the climax stage in different time spans. A botanist or ecologist can examine the community composition and judge how long ago the site was disturbed and approximately how long it will take until the climax community will be reached. For example, grasslands take approximately 20 to 40 years, Eastern deciduous forests 100 years, and tundra many hundreds of years to reach the climax stage after a disturbance occurs.

Community development, or ecosystem change, occurs in aquatic, as well as terrestrial, ecosystems. The change in aquatic systems is caused by the process of **natural eutrophication**. Natural eutrophication is the increase of nutrients in aquatic systems because of natural community activities.

When a lake is formed by the retreat of a glacier, it is only a depression filled with clean, nutrient-poor water. It contains no sediments or living organisms. Over time, pioneer autotrophs and heterotrophs populate the new lake. When they die, they fall to the bottom. Decomposers break down the organic material, adding nutrients to the lake. The lake also receives runoff from the surrounding watershed, adding soil and other organic materials, further increasing the lake's nutrient level. As the nutrient concentration increases, the productivity of the lake increases. Over time, sediments build up and the lake becomes shallower and richer. Given enough time, the lake fills up and becomes a marsh. Finally, the marsh dries out and becomes part of the surrounding terrestrial ecosystem. Depending on the original diameter and depth of the lake and the rate of natural eutrophication, the lake can be short- or long-lived. The rate of the aging of a lake can be increased by cultural eutrophication. (See the nitrogen cycle.) The stages of succession in an aquatic environment are sequential and predictable under normal conditions.

Ecological succession is linear in most cases, extending from the pioneer stage to the climax stage. However, some ecosystems do not exhibit linear progression, but instead have a cyclic sequence. **Cyclic succession** means that the community goes through several seral stages, is disturbed by some natural disaster, and starts over again. The community is set back to the pioneer stage by some naturally reoccurring event, so that the system never reaches the climax stage for that physical environment.

Fire is a common cause of cyclic succession. The pinelands (Pine Barrens) of southern New Jersey and the chaparral of southern California display the results of periodic fires caused by lightning strikes. If there were no fires, the pinelands and chaparral would only be seral stages, not climax stages, for those areas. Fire is an important force in maintaining not only fire climax communities, but also linear climax communities. For example, in a prairie where the climax is a grass community, fire is a normal event. Without periodic fires, the trees from the surrounding communities would move out into the prairie, replacing the prairie with a forest. The fire kills off the young tree seedlings that grow out into the grass community.

Human activity can alter this natural succession. For example, the U.S. Forest Service devised a policy to eliminate forest fires: Smokey the Bear. Many national forests are used heavily for human recreation, and the Forest Service has had a campaign for years to stop all forest fires. With no fires, dry underbrush (litter) accumulates on the soil surface around the base of trees. When there finally is a fire, the accumulated underbrush burns hotter and longer than a natural fire would.

A forest fire, when there is little litter build-up, is a low temperature fire (**ground fire**) that moves quickly through the forest at the ground level. A ground fire does not damage mature trees. However, if a fire occurs where there is a large amount of litter, it burns hotter and longer (**crown fire**). A crown fire climbs to the tops of the trees, killing them. In recent years, the Forest Service has started experimenting with controlled burns to reduce the build-up of undergrowth and, thereby, the possibility of crown fires. An interesting note is that many forest pine cones actually are adapted to ground fire. They do not release their seeds unless there has been a ground fire. However, a hotter crown fire can destroy the cones and their seeds. Again, it is important to fully understand a system before trying to alter it to suit human needs.

ASSIGNMENTS FOR PART 2: ECOSYSTEMS
HOW DOES THE ECOSYSTEM CHANGE?

1. Explain to a friend how succession works.
2. Compare a pioneer community to a climax community.
3. Explain to a friend the differences between primary and secondary succession.
4. Draw a graphic representation of the life cycle of a lake (birth to death).
5. Explain natural eutrophication to a friend.
6. Use an example of your choice to explain cyclic succession.

SUGGESTED READINGS

Krebs, Charles. (1994). *Ecology* (4th ed.). New York: Harper-Collins. (ecology textbook)

Starr, Cecie, & Taggart, R. (1995). *Biology: The unity and diversity of life* (7th ed.). New York: Wadsworth Publishing Company. (biology textbook)

BIOSPHERIC PROBLEMS

Human activities have led to environmental disruptions and biospheric problems. You have read about several biospheric problems and their causes (e.g., increased soil loss [erosion], water pollution, air pollution, acid rain, ozone layer depletion, increased greenhouse gases, global warming, biological magnification, overpopulation, nutrient cycle disruptions, cultural eutrophication, overharvesting in terrestrial and aquatic systems, salinization of desert systems, toxic waste). These have resulted in the degradation of the biosphere and habitat destruction. All of the above factors contribute to the loss of **biodiversity**.

Biodiversity is the total of all plants, animals, fungi, and microorganisms in the biosphere or in a specified area. It is estimated that there are from 10 to 100 million species in the world today, of which only 1.4 million have been identified and named. (See Table 2-25.) The vast majority of species has not been identified (e.g., only 69,000 fungi have been named out of an estimated 1.5 million species).

Table 2-25: **Examples of Numbers of Identified Species**

Group	Number of Species
Birds	9,000
Mammals	4,500
Butterflies	20,000
Plants	250,000
Fungi	69,000
Bacteria	4,000

Not every species that has ever evolved on earth is still in existence because species naturally die out, or become extinct. A species can live anywhere from one to 100 million years. Over the fossil record, the number of species gained by the process of speciation has been greater than the number of species lost through extinction. Therefore, the number of species has increased over time. The normal rate of extinction is estimated at 12 to 20 species per year. This rate has remained unchanged for the last 65 million years. Most of the extinctions since prehistoric times have occurred in the last 300 years, and most of these have occurred in the last 50 years. If we look only at the last 50 years, most of the extinctions have occurred in the last 10 years.

During the second half of the twentieth century, the rate of extinction has increased more than 1,000 times. Some scientists estimate that the extinction rate is now 25,000 times the natural rate, and that by 2050, more than half of all remaining species will have vanished. It is estimated that by the year 2000, one million species will be lost and 15 to 20% of all plant species will be gone because of human impact and environmental degradation. The rate of extinction is about 100 species each day.

The threat to global biodiversity increases yearly. Of 9,000 bird species, 3,000 are holding their own and 6,000 are in decline. Of the 270 turtle species, 42% are threatened with extinction. In the mid-Atlantic United States, between the 1940s and the 1980s, migratory songbird populations decreased 50%. Twenty percent of the world's freshwater fish species are extinct or in dangerous declines. Of all native North American animals, 33% of fish species, 11% of bird species, and 13% of mammal species are rare or extinct.

Why is there a sudden increase in species extinction? One of the main causes of species extinction, or loss of biodiversity, is the loss of habitat. Biodiversity is logarithmically related to the size of an area. This means that an area of 10 square miles will hold twice as many species as an area of one square mile. If you reduce an area to one-tenth its original size, half of the organisms will become extinct in that area.

Species exist in complex ecological system. If a habitat is reduced, species extinction can occur even if the entire habitat is not destroyed. Decreasing the size of, or dividing, a habitat is called **fragmentation**. It is not enough to save small pieces of habitat to ensure biodiversity. The habitat must be large enough for the various species to have all of their needs met.

Ninety percent of all biospheric species live in terrestrial habitats. Of these, two-thirds to three-fourths live in the tropics. More than half of these tropical species live in the tropical rainforests. It is expected that by 2025, at current rates, 90% of all tropical rainforests that now exist will be gone and, along with them, 50% of the species that live in them. Because more than half of all species on earth live in tropical rainforests, one-quarter of the earth's species will become extinct within your lifetime. Most of these species are unknown. Each year an area of tropical rainforest the size of Nebraska is destroyed.

There are many causes of extinction. (See Table 2-26.) Because habitat loss and predator and pest control have been discussed previously, I will now discuss some of the other causes of extinction. Many species are endangered today because of the practices of using animals for medical research, collecting for individuals and zoos, and owning pets. Compared to deforestation, these practices may sound trivial in their effect on global biodiversity, but they do have a significant impact on certain groups of species. For example, in 1980, more than 128 million tropical fish, two million reptiles, and one million other wild animals were sold in the United States as pets. Many of these are rare or endangered species.

Table 2-26: Some Causes of Species Extinction

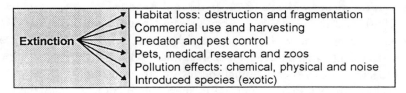

Collectors and gardeners also affect plants species. One-third of the cactus species native to the United States are now endangered because so many were collected and sold as potted plants. Nine species of birds are now threatened or extinct because of collection and the pet market. Endangered butterflies are being driven to extinction because so many have been captured for butterfly collectors. Tropical reefs, some of the most productive ecosystems in the world, are being damaged and killed by collectors of tropical fish for the pet markets of the world.

Commercial use and harvesting has been discussed in terms of deforestation and overfishing, but another example of habitat destruction and the consequent loss of biodiversity is the mining of coral in the Indian Ocean. Coral is pulverized and used to make cement. In Sri Lanka, where the harvesting of coral is illegal, approximately 10,000 tons of coral are removed each year for the construction industry. That is enough coral to fill 180 railway cars. It is estimated that about 80% of the country's reefs have been damaged, resulting in the loss of natural water breaks for the land and destruction of nursery habitat for hundreds of marine plants and animals. At the tip of India, in the Gulf of Mannar, 40,000 tons of coral are mined each year.

The effects of pollution, both chemical and physical, and how it alters the environment so that an organism's tolerance limits are exceeded, have been discussed in this text. Another kind of pollution that often goes unnoticed is noise pollution. We know that exposure to loud sounds can cause hearing loss, nervous disorders, and psychological and physiological stress in humans. Noise pollution affects other species in similar ways, notably marine mammals. Under natural conditions, the ocean is a quiet place, but underwater noise has increased with human activity. (See Table 2-27.) Marine mammals, such as whales, seals, and sea otters, use sound to communicate, find food, and migrate. Increased noise pollution can disrupt these activities and, if loud and long enough, it may cause deafness. It has been estimated that whales once communicated over a distance of 1,000 miles, but that with increased ocean noise pollution, this range has decreased to 500 miles.

Table 2-27: Sources of Noise Pollution in the Oceans

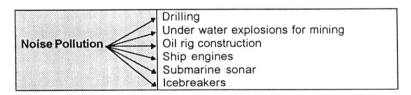

Recently, a scientific experiment was proposed that would help scientists gather data on global climate change. This experiment would measure the temperature of the ocean by measuring how fast sound travels through the deep ocean. Water temperature affects the rate of speed at which sound travels. To run the experiment, underwater speakers would send sound waves across the Pacific Ocean at least once a day, every day, for two years. This increase in noise pollution raised questions about possible effects on marine mammals.

It is possible that the extra noise pollution could deafen these mammals. They then would not be able to communicate, find food, or navigate. Once again, the precautionary principle needs to be taken into account.

Another disruption to an area that is often overlooked is introduced species. These species can directly cause extinction and endangerment. Because an introduced species has not evolved in an area, it often has no local predator, and so can reproduce freely. Frequently, introduced species out-compete native species for nesting sites, territory, and food; cause habitat damage; and prey on native species. Of the fish on the U.S. endangered species list, half are threatened by introduced species. Two-thirds of extinct fish species became extinct because of introduced species. (See Table 2-28 for examples of introduced species and their effects.)

Table 2-28: Examples of Introduced Species and the Damage They Cause

Name	Damage
European starling	Compete with native songbirds, crop damage
House sparrow	Damage crops, compete with native songbirds, occupy bluebird nesting sites
Walking catfish	Destroy bass, bluegills, and other fish
Carp	Compete with and displace native fish, destroy habitat, decrease water fowl populations
Japanese beetle	Defoliate trees and crops
Water hyacinth	Clog waterways, out-compete other native vegetation
Loosestrife	Compete with and displace native wetland plants
Chestnut blight (fungus)	Has destroyed almost all eastern American chestnut trees
Dutch elm disease (fungus)	Destroys elm trees

Source: Living in the Environment by G. Tyler Miller. ©1985 by Wadsworth Publishing Co.

The introduction of nonnative species (exotics) can be accidental or intentional. It is difficult to contain an introduced species, and once it escapes into the environment, the species is almost impossible to control. For example, rabbits were introduced to Australia as a food source, but some escaped into the wild, causing widespread habitat destruction. The African "killer" bee was imported into South America to increase honey production. However, some of these aggressive bees escaped and have since spread north as far as the southern United States.

The introduction of a new species to an area is often intentional (e.g., the introduction of pigs, cats, and dogs to the Galapagos Islands; pigs to the Hawaiian Islands; game fish into the American water system). The effects of these introductions have been disastrous for many native species. The Galapagos tortoise has become endangered because their eggs and their young are preyed upon, songbirds in Hawaii are declining because their habitat is being destroyed, and native American fish are out-competed or preyed upon by these introduced species. (See Table 2-29 for the effects of introduction on some native species.) As you can see, the introduction of a nonnative species can have widespread repercussions.

Table 2-29: Effect of Stocking of Game Fish in the United States

Game Fish	Affected Area	Native Species
Brown trout	California	golden trout
Trout, catfish	Arizona	trout, catfish
Green sunfish	Texas	Comanche Springs pupfish
Pacific salmon	Great Lakes	lake trout
Largemouth and smallmouth bass	Maine	Atlantic salmon

Ecosystems are complex and interconnected. The loss of a single species can wide-ranging effects on a system. The fruit bat plays a central role in tropical ecosystems. It is the principle pollinator and disseminator of seeds for many plants. The decline of the fruit bat has resulted in the decline of fig production, and because figs are a key producer in the tropical system, the whole system suffers.

How can the loss of biological diversity and species extinction be stopped or slowed? One way is to stop habitat destruction and fragmentation. To date, the most effective way to accomplish this has been the Endangered Species Act of 1973. This legislation sets forth the procedures for protecting plants and animals that are in danger of becoming extinct. The way to protect an endangered species is to save or protect its habitat.

The success of the Endangered Species Act can be illustrated by the California gray whale. Annually, California gray whales make a round-trip migration from Alaska to Baja California Sur. The gray whale had been hunted nearly to extinction by the 1950s, and was placed on the Endangered Species List. This resulted in a cessation of hunting and in efforts to protect the whales' habitat, especially their breeding grounds.

In 1988, the Mexican government set up the Vizcaineo Biological Reserve in Baja California Sur to protect the whale's breeding and calving grounds. The sanctuary includes the three lagoons where California gray whales migrate each winter to mate, reproduce, and raise their calves. The calving lagoons are highly saline, which makes the whales more buoyant and calving easier. The most pristine lagoon in the Vizcaineo Biological Reserve is Laguna San Ignacio, which is two miles wide and 20 miles long. Approximately 500 mature whales are in residence there during the breeding and calving season. Because of these efforts, the whales' population reached 20,000. This plan proved to be such a success that, in June 1994, the California gray whale was the first marine mammal to be removed from the Endangered Species List.

Then, in July 1994, a proposal was made to build a salt production plant in the San Ignacio Lagoon. The plant would remove 800 million metric tons of water per year, producing six million tons of salt annually, or 16,000 metric tons daily. The operation would cover 130,000 acres of land, one entire shore of the lagoon. A mile-long pier jutting out into the Pacific Ocean would be constructed for loading and transporting the salt.

How would this affect the whale? No one knows how a pier extending out into the ocean would affect the whales' migratory patterns; how the increased human activity in and around the lagoon would affect their breeding and calving; or how the reduced salinity, and consequent reduced buoyancy, would affect the whales' calving. Once again, the precautionary principle must come into play. The Endangered Species Act is currently the only powerful weapon available to save species by saving their habitat. Finding a way to balance commercial and industrial development, agribusiness needs, and suburban sprawl — and their resulting habitat destruction and fragmentation — with the need to maintain biodiversity is a difficult job.

Biodiversity is vital for providing food security, for preserving a global gene bank to ensure genetic variability for disease and pest resistance, and for supplying a source of new prescription drugs and solar-powered productivity. Forty percent of prescription drugs come from wild plants, animals, and microorganisms, and 80% of herbal medicines come from forests. Once a species is extinct, we will never know what it had to offer us or the part it played in keeping the local or global ecosystem running. It is impossible to measure the value of an ecosystem or a species.

Recently, "risk assessment" and cost-benefit analysis have been suggested as ways to decide whether to save a species, protect a habitat, or stop a source of pollution. This approach weighs a species' worth against how much it would cost to maintain it, or how much a habitat is worth to the biosphere.

But how do you measure the environmental benefit or worth of clean air, clean water, a mature forest, or the gray whale? You can't quantify nature's benefits; moral and ethical arguments cannot be discussed in terms of dollars. The tools used in cost-benefit analysis are not exact, have large margins of error, and are dependent on many assumptions and value judgments. They are not free of personal bias. Every living thing has intrinsic value, and we must find ways to preserve habitat and life. We must find a balance between human welfare and economic growth and protecting physical resources and preserving the biodiversity of the biosphere.

ASSIGNMENTS FOR PART 2: ECOSYSTEM
BIOSPHERIC PROBLEMS

1. Explain to a friend what biodiversity is and how it relates to the size of an area.
2. Construct a graph that shows the natural rate of extinction and the current rate of extinction.
3. Explain to a friend why the rate of extinction has increased over the last 50 years.
4. Why is the size of a habitat important to species survival?
5. Explain to a friend what the Endangered Species Act is and why it is important.

6. What is biodiversity and why is it important?
7. Explain to a friend how introduced species can affect an ecosystem.
8. Go to the library and find two articles on introduced species and their effects on native species or systems. Summarize what you learned from the articles.
9. Using a specific example of your choice, explain why you would prefer to use risk assessment and cost-benefit methods in a debate on air quality.
10. Explain to a friend why using risk assessment and cost-benefit analysis to decide an environmental is not a good idea.
11. Why was the precautionary principle important when the underwater experiment on sound was discussed?

SUGGESTED READINGS

Goldfarb, Theodore D. (1993; yearly). Taking sides: Clashing views on controversial environmental issues. Guilford, CT: The Dushkin Publishing Group, Inc. (pros and cons on environmental issues)

Miller, G.T., Jr. (1994). Living in the environment (8th ed.). Belmont, CA: Wadsworth Publishing Co. (environmental science textbook with case studies)

Saur, K.G. (1994). Encyclopedia of world problems and human potential (4th ed.). New York: K.G. Saur; R.R. Bowker Co. (environmental issues and human abilities)

Wilson, Edward. (1992). The diversity of life. New York: W.W. Norton & Co., Inc. (issues of biodiversity and the environment)

Worldwatch Institute. The state of the world 1994-1995. New York: W.W. Norton & Co. (annual update of current world environmental issues and what needs to be done)

Worldwatch Institute. (1995). Vital signs: The trends that are shaping our planet. New York: W.W. Norton Co. (annual update showing global trends of environmental concern, published annually)

PART 2: BIOMES

WHAT ARE THEY?

If you have traveled across the United States or around the world you have probably noticed certain areas that are quite far from one another look similar. For example, the chaparral of southern California looks like scrubland areas of the Mediterranean, the forests of the northeastern United States resemble the forests of England, and the evergreen forests across the United States resemble one another. My grandfather used to say that he was always disappointed when he traveled outside the United States because everywhere he traveled, the natural areas looked like someplace in the United States. He did not say that they looked, or were, identical, but that they were similar.

If someone asks you what a desert looks like, you can describe it pretty accurately, whether it is in Africa, South America, Asia, or the United States, because all deserts have similar characteristics (e.g., they are dry, have no large trees, the vegetation is scattered, and there are no herds of large herbivores). The community compositions in the Sonoran and Gobi deserts are different, but they look similar because the species have evolved in response to similar environmental pressures.

Why is it that areas that have similar climates produce plant and animal associations that resemble one another? They have similar limiting factors, so they shape similar communities. Large, complex, stable global ecological communities with distinct and recognizable associations of plants and animals are called **biomes**. Although altitude and soil conditions influence each biome, its community is primarily determined by the regional climate (i.e., rainfall and average seasonal temperatures).

Scientists who study biomes have composed different groupings, organizations, and subdivisions. Aquatic biomes are divided into two major classes: marine and freshwater. Terrestrial biomes are divided into three major classes: forests, grasslands, and deserts. These classes can be further divided into tropic, temperate, taiga, and tundra. For detailed information about a biome, read several ecology texts. Each author covers biomes in a different fashion: Some emphasize animals, some plants, some give only general climatic information. (See Table 2-30 for a list of aquatic and terrestrial biomes and their major classes and subdivisions.)

Table 2-30: List of the Major Aquatic and Terrestrial Biomes

AQUATIC	
	Marine
	oceans
	estuaries
	saltwater marshes
	Freshwater
	streams and rivers
	ponds and lakes
	freshwater marshes
TERRESTRIAL	
	Tropical rainforests
	Tropical seasonal (deciduous) forests
	Tropical thron forests
	Savannas
	Temperate deciduous forests
	Woodland and shrubland (chaparral)
	Grassland (prairies)
	Deserts
	hot
	cold
	Taiga (boreal)
	(northern evergreen)
	(coniferous) forest
	Tundra
	alpine
	arctic

MAJOR BIOMES IN THE UNITED STATES

Oceans cover approximately 70% of the earth's surface. It is the thickest ecosystem and the most diverse on the planet. Oceans moderate land climates and the amount of carbon dioxide and oxygen in the atmosphere. Physical factors (i.e., waves, tides, currents, temperature, pressure, salinity, and light intensity), dominate the life in this biome and determine the community composition. As one goes from the topmost layer of the ocean (the eutrophic zone) to the lowest level, the amount of light and dissolved gases, and the temperature decrease. Producers are limited to the thin upper layer (eutrophic zone) and form the energy base for this biome. The ocean is divided into zones: the continental shelf, continental slope, continental rise, and the lowest level (the abyssal plain). Most human activities are concentrated on the continental shelf (e.g., mining for minerals, drilling for oil, harvesting seafood).

Estuaries are semi-enclosed bodies of water in which freshwater and saltwater meet (e.g., at the mouths of rivers coastal bays). They have daily fluctuations in salinity because of tidal activity, and they are one of the most fertile and productive biomes. They act as nurseries for many marine organisms.

Freshwater streams and **rivers** are among the biomes that are most intensely used by man. They are unique because they change characteristics from their source to their mouth, or where they empty into the ocean. Their size and volume of water, as well as their species composition, changes along their length. Human impact has been high on these biomes (e.g., soil erosion, the dumping of waste materials, damming, fertilizer runoff, the introduction of nonnative species for sport).

Freshwater ponds and **lakes** are standing freshwater. The size, depth, and chemical characteristics of these biomes change annually and over geologic time. Artificial or cultural eutrophication has accelerated their "aging." The introduction of nonnative species for sport, pollution caused by fertilizers, lake community runoff, fossil fuel from boats, and acid rain have all damaged these biomes.

Freshwater marshes are naturally fertile biomes. They experience changes in water level because of annual rainfall, and they are dependent on periodic fires for their maintenance. They are important as duck and fur animal habitats, and they help maintain the water table of adjacent ecosystems.

Terrestrial biomes are divided into subclasses based on temperature, the amount of annual rainfall, and its seasonal distribution. **Deserts** are some of the driest biomes, receiving less than 10 inches (25 cm) of precipitation annually. There are two kinds of deserts: hot and cold. In the United States, we have both types (e.g., a cold desert in Washington state and hot deserts in the Southwest). Desert plants and animals have adapted to cope with the lack of water. Desert animals are primarily nocturnal, or they are reptiles. Desert plants grow only when there is enough water to ensure reproduction (annuals), have long root systems to reach the low water table, store water in plant parts, go dormant during droughts by dropping their leaves, and produce chemicals into the soil that establish territories in order to decrease competition for water.

Tundra biomes are cold and dry, with an annual precipitation of between four and 10 inches (10-25 cm). They are limited by temperature, with summer temperatures never below about 50° F. Tundras are treeless and the vegetation is composed mostly of lichens, grasses, and sedges. Tundra animals are generally migratory (e.g., reindeer, musk ox, polar bears). This system is very delicate because of its short growing season and the slow build-up of soil. Most of the nutrients are locked up in this biotic community.

Grasslands or **prairie** biomes have cold winters and receive between 10 and 30 inches (30-85 cm) of precipitation annually. The producers are grasses and the biomes are maintained by seasonal droughts and periodic fires. Prairie heterotrophs are primarily large grazing mammals (browsers) (e.g., bison, antelope, kangaroos). Because of the lack of cover, the animals are those that run or burrow to escape predation. These are very productive biomes, and large amounts of organic matter accumulate each year, producing rich deep soils. Their high productivity has led humans to overgraze and overplow, or farm them.

The several types of **forest** biomes are determined by moisture and temperature. The northern evergreen forests (the taiga, boreal forests, and coniferous forests) are found at cool, dry, and high elevations. They have long, cold winters and short, dry summers. They receive between eight and 40 inches (40-100 cm) of precipitation annually. Their trees are evergreens, dropping one-fourth to one-third of their leaves annually, and they reproduce by forming cones (gymnosperms). The soils are thin and nutrient poor. Deciduous forests receive between 30 and 60 inches (70-140 cm) of precipitation each year, which is evenly distributed throughout the year. The temperature during the growing season is 60-65° F. Deciduous trees reproduce by forming flowers and fruits (angiosperms) and lose all of their leaves each year. Tropical forests are found in warmer, wetter regions. The mean temperature is 82° F and the average annual precipitation is 100 inches (250 cm). Tropical rainforests are some of the most productive biomes of the biosphere. Their soils are nutrient poor because most of the nutrients are locked up in the standing biomass.

The **shrubland** (**chaparral**) biome occurs in areas that experience mild, rainy winters and long, hot, dry summers. The low annual precipitation falls in the winter months. This biome is maintained by fire.

Although similar biomes have many characteristics in common (e.g., the Sonoran and Gobi deserts), a complete understanding the structure and function of a specific biome requires individual research.

ASSIGNMENTS FOR PART 2: BIOMES

1. In your own words, describe to a friend each of the biomes listed below (i.e., what kinds of plants, animals and climate you would expect and how each looks).
   a. oceans
   b. estuaries
   c. saltwater marshes
   d. freshwater (streams, lakes, and marshes)
   e. deserts
   f. tundra
   g. grasslands (prairies)
   h. forests (evergreen, deciduous, and tropical)
   i. chaparral
2. This assignment will allow you to synthesize and apply all of your knowledge about ecological principles and biospheric problems to a specific biome. For each of the biomes listed below, give at least three ways human activity has affected it.

   Example: The following is a partial list for Oceans:
   oil drilling and tanker accidents (pollution)
   offshore dumping (pollution)
   overharvesting (decrease in biodiversity)
   mining of minerals (habitat destruction)
   depletion of the ozone layer (decreased productivity)

After you make your list, explain to a friend how each activity affects the biome using specific examples.

    a. Oceans

    b. Estuaries

    c. Freshwater

        1. streams

        2. lakes

    d. Marshes

    e. Deserts

    f. Tundra

    g. Grasslands (prairie):

    h. Forests

        1. evergreen

        2. deciduous

        3. tropical

    i. Chaparral

3. Which biomes are fire maintained? Explain how fire maintains them.

4. Which biomes are the most productive and why?

5. Which biomes have nutrient-poor soil and why?

SUGGESTED READINGS

Miller, G.T., Jr. (1994). *Living in the environment* (8th ed.). Belmont, CA: Wadsworth Publishing Co. (environmental science textbook with case studies)

Starr, Cecie, & Taggart, R. (1995). *Biology: The unity and diversity of life* (7th ed.). New York: Wadsworth Publishing Company. (biology textbook)

PART 3: WHAT YOU DO MAKES A DIFFERENCE

This book has introduced you to some basic ecological principles, the structure and function of ecosystems, and ways to approach biospheric issues. You know that the information given by the television announcers, politicians, and newspaper reporters is generally incomplete and biased. With the information in this book, you now have the background to ask the questions that will get you the information that *you* need to make decisions about issues that deal with the biosphere. (See Appendix I.)

When you consider solutions or approaches to a problem, remember that you need to differentiate between short-term and long-term responses. You should recognize the biospheric impact of what is going on around you, the trade-offs of different solutions, and the consequences of waiting to respond to a problem until all of the "information" is in.

This book is just a beginning. Remember that the biosphere, like a spider's web, is a system in which everything is interconnected and the least alteration affects the whole. The holistic view is essential to understanding the structure and function of any system. Everywhere you turn today you see global ecological decline. The biosphere is being degraded as a result of human activities, and unless there is a rapid response to this degradation, the biosphere will cease to function in a life-sustaining fashion.

Since we don't know how to rebuild the systems of the biosphere, we have to do everything we can to stop this degradation. This means considering not just individuals' rights, but also the rights of the biosphere and of all of its living components. You have a responsibility not just to yourself, but to all of life on earth to consider the effects of your actions, or your inaction. Your personal choices, whether they include conserving resources by using less, buying environmentally safe products, voting for legislation that protects the biosphere, or changing your diet, impact the environment, and, ultimately, the whole world. Everything you do either helps or hurts.

Some people say that there is nothing to worry about because things are getting better. It is true that some things are better now than they were 25 years ago. The question is: Why are they better? The air and water are cleaner in some areas because local, state, and federal laws have been passed that protect the environment. These laws have set standards so that environmental pollution and damage is curtailed.

However, if these laws are weakened or removed there are no incentives for polluters to make decisions for the good of the environment. The majority of businesses and individuals do not see that the environment is their responsibility. They only see what they want to do, how much money they can make, or what is in their best interest. It is cheaper in the short term to pollute and overharvest than not to pollute and conserve. The biosphere and you are the ones who will suffer. Only you and the laws that are passed by the people you elect can protect the earth. How you vote affects local, state, and federal activities, laws, and decisions. A healthy environment and biosphere cost money, time, effort, personal involvement, and commitment. If we take care of the biosphere, it *will* take care of us.

Because we cannot make a functioning system, we have to protect the one we have. We must clean up and stop pollution and environmental degradation. The precautionary principle should be considered before adopting any proposal.

The main causes of environmental degradation and the problems in the biosphere can be traced directly to human overpopulation and consumerism. The efforts to stop environmental degradation must be individual, as well as national. As you have seen, every choice you make has a consequence, from buying a fuel-efficient car to choosing cereal instead of bacon for breakfast to turning off the water when you brush your teeth — to voting. Through the cooperation and coordination of people like you all over the world, biospheric change is possible. What happens in China affects you as surely as what happens down the block or in the next town. You can act locally, but you always need to be aware of global activities that affect the biosphere.

It is my hope that your awareness of the biosphere, how it works, the challenges it faces, and how you can make a difference have been increased by reading this book. I hope that this book is only the beginning for you. Change requires conscious effort by all of us. You can, and you do, make a difference.

ASSIGNMENTS FOR PART 3:
WHAT YOU DO MAKES A DIFFERENCE

1. Design a detailed life action plan for yourself that will benefit the biosphere.
2. Implement your plan.

SUGGESTED READINGS

Abbey, Edward. (1968). *Desert solitaire.* New York: Ballantine Books. (or any other books of his naturalist essays and the philosophy of environmental activism)

The Earthwork Group. (1989). *Fifty simple things you can do to save the earth.* New York: Harper Collins. (booklet that describes things that you can do in your daily life to save the earth)

The Earthwork Group. (1991). *A student environmental action guide.* New York: Harper Collins. (list of things that you can do as a student to change the earth)

Eiseley, Loren. (1978). The star thrower. In Loren Eiseley, *The star thrower.* New York: Times Books. (naturalist essay on man's relationship to nature and individual responsibility)

Stein, Sara. (1993). *Noah's garden: Restoring the ecology of our own backyards.* New York: Houghton Mifflin Co. (how the backyard garden works as an ecosystem)

True understanding in an individual has two attributes: awareness and action.

Who can enjoy enlightenment and remain indifferent to the suffering in the world?

Only those who increase their service along with their understanding can be called men and women of the way.

Source: Brian Walker. *Hua Hu Ching: The Unknown Teachings of Lao Tzu.* 1994, Harper Collins, revised edition.

APPENDIX I

APPROACH TO BIOSPHERIC ISSUES

By the end of the semester, when you read an article in the newspaper or watch a program on TV concerning a biosphere or some other environmental issue, you should be able to answer the following questions:

1. What is the issue, question, or problem under discussion?
2. What are the scientific laws, theories, principles, concepts, or processes that are important to understanding the issue?
3. What are the arguments for continuing the current practices?
4. What are the arguments for changing or discontinuing the current practices?
5. What are the possible solutions? What are the ecological and economic short- and long-term effects of each solution?
6. Which solution do you prefer? Explain your reasoning.

This book discusses the following scientific theories, laws, principles, concepts, and processes:

biogeochemical cycling	law of entropy
biological magnification	modeling
biomes	modern view
carrying capacity	photosynthesis
cellular respiration	precautionary principle
ecosystem succession	principle of limiting factors (law of
energy flow	tolerance)
food chains	scientific method
food webs	soil development and erosion
Gaia hypothesis	species diversity
Gause's principle	theory of evolution
law of conservation	trophic classification

APPENDIX II

COMMON EQUIVALENT MEASUREMENTS

Length
    Metric
    1 kilometer (km) = 1,000 meters (m)
    1 meter (m) = 100 centimeters (cm)
    1 meter (m) = 1,000 millimeters (mm)
    1 centimeters (cm) = 0.01 meter (m)
    1 millimeter (mm) = 0.001 meter (m)

    U.S.
    1 foot (ft) = 12 inches (in)
    1 yard (yd) = 3 feet (ft)
    1 mile (mi) = 5,280 feet (ft)

    Metric-U.S.
    1 kilometer (km) = 0.621 mile (mi)
    1 meter (m) = 39.4 inches (in.)
    1 inch (in) = 2.54 centimeters (cm)
    1 foot (ft) = 30 centimeters (cm) = 0.305 meters (m)
    1 yard (yd) = 0.914 meter (m)
    1 mile (mi) = 1.61 kilometers (km)

Area
    Metric
    1 hectare (ha) = 10,000 square meters (m^2)
    1 hectare (ha) = 0.01 square kilometers (km^2)

    U.S.
    1 acre (ac) = 43,560 square feet (ft^2)

Metric-U.S.
1 hectare (ha) = 2.471 acres (ac)

Volume
Metric
1 liter (L) = 1,000 milliliters (mL) = 1,000 cubic centimeters (cm³)

U.S.
1 gallon (gal) = 4 quarts (qt)
1 quart (qt) = 2 pints (pt)

Metric-U.S.
1 liter (L) = 0.265 gallon (gal)
1 liter (L) = 1.06 quarts (qt)
1 barrel (bbl) = 159 liters (L)
1 barrel (bbl) = 42 U.S. gallons (gal)

Mass
U.S.
1 ton (t) = 2,000 pounds (lb)
1 pound (lb) = 16 ounces (oz)

Metric-U.S.
1 metric ton (mt) = 2,200 pounds (lb)
1 kilogram (kg) = 2.20 pounds (lb)
1 pound (lb) = 454 grams (g)
1 gram (g) = 0.035 ounce (oz)

Energy and Power
Metric
1 kilocalorie (kcal) = 1,000 calories (cal)
1 calorie (cal) = 4.184 joules (J)

Metric-U.S.
1 kilojoule (kJ) = 0.949 British thermal units (Btu)
1 kilowatt-hour (kW-h) = 3,400 British thermal units (Btu)

Approximate crude oil equivalent
   1 barrel (bbl) crude oil = 6,000,000 British thermal units (Btu)
   1 barrel (bbl) crude oil = 2,000 kilowatt-hours (kW-h)

Approximate natural gas equivalent
   1 cubic foot (ft3) natural gas = 1,000 British thermal units (Btu)
   1 cubic foot (ft3) natural gas = 0.3 kilowatt-hour (kW-h)

Approximate hard coal equivalent
   1 ton (t) coal = 20,000,000 British thermal units (Btu)
   1 ton (t) coal = 6,000 kilowatt-hours (kW-h)

Temperature Conversions
   Fahrenheit (°F) to Celsius (°C):
      $°C = 5/9(°F - 32)$ or $°C = \dfrac{(°F - 32.0)}{1.80}$
   Celsius (°C) to Fahrenheit(°F)
      $°F = 9/5(°C) + 32$ or $°F = (°C \times 1.80) + 32.0$

GLOSSARY

A

Abiotic: The nonliving environment.

Abiotic synthesis: The formation of organic molecules from nonliving materials by abiotic interactions.

Absorb: To take in.

Acid rain: Rain that contains sulfuric and nitric acid; rain that lowers the pH of terrestrial and aquatic ecosystems, stressing their biotic components and affecting the systems' abilities to function normally.

Acre: Unit of area equal to 43,560 square feet, or the size of a football field.

Active pool: *See exchange pool.*

Aggregate: The grouping of objects.

Agribusiness: The industrialization of farming; farming as cash crop production; to make the most money per acre by increasing productivity and decreasing production costs without regard to the effects on the ecosystem.

Agricultural industrialization: *See agribusiness.*

Air pollution: Unfavorable changing of the atmosphere because of human activity; chemicals in the air in sufficient concentration to harm any living organism or damage nonliving matter.

Anabolic processes: Processes that build up matter.

Applied science: Research done by scientists working in industry or for the government, the resulting information is not public knowledge; making things that are useful to mankind.

Aquaculture: The farming of fish and shellfish in land-based ponds or contained aquatic systems.

Aquifer Underground: basin of geologic water.

Atlantic Engine: The interaction of wind and ocean currents in the North Atlantic that remove excess carbon dioxide from the atmosphere.

Atmosphere: Region of gases and particulate matter extending above the earth's surface.

Autotroph: Organism that makes its own food from inorganic molecules.

B

Bar graph: A graphing technique used to present discontinuous data.

Basic science: Research done by scientists in academic institutions, resulting in information openly published and accessible; knowledge for the sake of knowledge.

Bedrock: Unmodified parent material; forms the R horizon of the soil profile.

Biodegradable: Capable of being broken down by bacteria into basic elements or compounds.

Biodiversity: The number of different species that lives in a system.

Biogeochemical cycle: Pathway by which a chemical element moves from the abiotic environment to the biotic environment, and back into the abiotic environment.

Biological control: The use of natural predators to control pest populations.

Biological magnification: The accumulation of materials as they progress through a food chain.

Biology: The study of life in all of its manifestations.

Biomass pyramid: Measures the amount of living matter at each trophic level and presents the information as total dry weight.

Biome: Large, complex, stable global ecological community with a distinct and recognizable association of plants and animals; adapted to a particular environment.

Biont: The first living single celled organism on earth.

Bioregenerative: The idea that plants, animals, and microorganisms generate materials, recycle materials, and control life necessities, and are responsible for maintaining the biosphere's life-support systems.

Biosphere: All of the earth's ecosystems and communities on a global scale.

Biotic: The living part of the environment.

Buffer strip: A strip of vegetation planted to reduce wind speed and soil loss.

C

C$_3$: The type of photosynthesis in which the first stable compound has three carbons.

C$_4$: The type of photosynthesis in which the first stable compound has four carbons.

Calorie: The amount of heat necessary to raise one gram or one milliliter of water one degree centigrade (C) starting at 15° C.

Capillary water: Water trapped and held by soil.

Carnivore: An organism that obtains its energy by eating other animals, either herbivores or other carnivores.

Carrying capacity *(K)*: The number of individuals that a system can support without damaging the resources of that system.

Catabolic process: Process that breaks down matter.

Causal hypothesis: The explanation of an observation by finding out what caused the observed event to occur.

Cellular respiration: The process by which high-energy organic compounds are broken down through a series of steps to release stored chemical bond energy that will do work for the organism.

Chlorophyll: A pigment that participates in the conversion of radiant energy to chemical bond energy.

Clear-cut method: Harvesting, or cutting down, all of the trees in an area.

Climax stage: The final stage in community development that is self-perpetuating and unchanging in its species composition.

Coacervate droplet: Nonliving aggregate of organic molecules that has some characteristics that resemble those found in living single-celled organisms.

Condition of existence: An environmental selective pressure to which a species must adapt.

Contour farming: Plowing that follows the contours of the land to reduce soil erosion.

Community: All the organisms of different species living in a designated area.

Community development: Community change over time in a predictable sequence of stages or phases.

Companion planting: Planting plants that naturally produce and give off chemicals that repel pests, or plants that attract insects away from a food crop.

Competitive exclusion principle: *See Gause's principle.*

Conclusion: A decision based on the analysis of the data.

Conservation tillage: Decreasing the amount of soil turnover.

Consumer: An heterotroph that gets its energy from eating other organisms.

Crop rotation: Annual rotation of crops; alternating crops that remove large amounts of nitrogen with crops that add nitrogen to the soil.

Crown fire: A hot fire that climbs to the forest canopy.

Cultivated system: A system altered by human activity, managed to increase productivity, and powered by solar energy and energy subsidies.

Cultural eutrophication: The increase of nutrients in an aquatic system because of human activity.

Cyclic succession: Community change in which the community goes through several seral stages, is disturbed by some natural disaster, and starts again.

Cycling pool: *See exchange pool.*

D

Data: Recorded information or observations.

Deforestation: The removal of forests.

Denitrifying bacteria: Bacteria that convert nitrogen compounds into their gaseous form.

Density-dependent population: *See S-shaped curve.*

Density-independent population: *See J-shaped curve.*

Detritus food chain: The transfer energy from dead organic matter by decomposer organisms.

Developed system: *See fabricated system.*

Differential reproduction: When more individuals in the next generation are adapted to the selective pressure in the environment because more adapted individuals lived to reproduce.

Directed creation: A theory that life was created by a supreme being.

E

Ecological indicator species: A species that is sensitive to changes in the environment, quickly reflects these changes, and helps ecologists assess the health of an ecosystem.

Ecological pyramid: Presentation of community structure based on the trophic classification.

Ecological succession: *See community development.*

Ecologist: An individual who studies ecosystems to understand how they are structured and how they function.

Ecology: The study of organisms (populations or communities), their interrelationships, and their interactions with the environment; the study of the structure and function of an ecosystem.

Ecosystem: An organized unit, including the biotic and abiotic parts of the unit, and their interactions.

Electromagnetic energy: The range of radiation.

Emergent property: Characteristic produced by the interaction of parts.

Energy: The ability to do work within a system.

Energy flow: The behavior of energy as it moves though a system.

Energy pyramid: Measures the rate of energy flow or the productivity at each trophic level.

Entropy: The disorder in a system.

Environmental pricing: The pricing of a product to reflect the cost of correcting the environment damage or lost ecological function done when harvesting and processing the material.

Environmental resistance: All factors that limit the maximum size of a population.

Equilibrium species: *See K-selected strategist.*

Erosion: Soil lost to an ecosystem because of actions of wind or water.

Essential nutrient: Element that is required for life.

Eutrophication: The natural nutrient enrichment of aquatic systems.

Exchange pool: The part of the nutrient cycle that contains a small amount of the element; the element moves rapidly between the biotic and the abiotic parts of the ecosystem; the element is in a form that is usable by organisms or is exchangeable.

Exchange rate: How quickly material moves between the reservoir and exchange pools.

F

Fabricated system: A man-made system powered by fossil fuel; it is energy intensive.

Feedback mechanism: Mechanism in which the end-product affects the process.

Fibrous root system: Plant has many roots of equal size close to the surface to the soil.

First law of thermodynamics: Energy can be transformed from one type of energy to another type of energy. During the transformation, no energy is destroyed and no new energy is formed, energy is conserved during energy transformations.

Food chain: The transfer of energy through a system from the producers through the saprovores, "who eats whom."

Food chain concentration: *See biological magnification.*

Food web: The pattern caused by interlocking food chains, a network of energy transfers within an ecosystem.

Fragmentation: Decreasing the size of, or dividing, a habitat.

G

Gaia hypothesis: Hypothesis proposed by James Lovelock, named after the Greek goddess for "Mother Earth"; the biosphere is self-regulating and the biotic parts of the biosphere actively interact to modify and control the chemical and physical conditions of the biosphere.

Gaseous cycle: A nutrient cycle which has its reservoir pool located in either the atmosphere or hydrosphere.

Gause's principle: No two organisms can occupy the same niche in an ecosystem indefinitely.

Generalist species: Species that have wide ranges of tolerance.

Ghost acreage: The acreage needed to grow food for countries that cannot grow enough to sustain their own populations.

Global circulation pattern: The movement of materials around the globe.

Global warming: Human-enhanced greenhouse effect; unnatural rise in the earth's atmospheric temperature because of increases in greenhouse gases.

Gradualist: *See neo-Darwinist.*

Grazer: *See herbivore.*

Grazing food chain: The transfer of energy from autotrophs to herbivores to carnivores.

Great conveyor: The circulating pathway of ocean currents around the globe.

Green pricing: *See environmental pricing.*

Green revolution: The introduction of new, high-yield crop varieties that need auxiliary energy in the form of fertilizer, water, and pest control to produce these high yields.

Greenhouse gas: Gas in the atmosphere that causes, or increases, the greenhouse effect.

Gross primary productivity: The total amount of carbohydrate fixed by photosynthesis in the system per area per day.

Ground fire: A low-temperature fire, fast moving at the ground level.

H

Habitat: Where an organism lives. *63*

Hectare (ha): Unit of area equal to 2-1/2 acres, or 2-1/2 football fields.

Herbivore: Organism that obtains its energy from eating producers.

Heterotroph: Organism that obtains its energy by eating other organisms.

Hierarchical organization: The nesting, or organization, of characteristics or units, with one smaller unit fitting into, or nesting into, the next larger unit.

Hierarchy: A graded series (e.g., going from smallest to largest or simplest to the most complex).

Holistic approach: The application of the modern view to the study of ecosystems. *See modern view.*

Homeostatic mechanism: Mechanism that maintains stability or equilibrium in a system.

Horizon: Layer in the soil profile with identifiable physical and chemical characteristics.

Hydrological cycle: The water cycle.

Hydroponics: Raising plants in nutrient water solution.

Hydropower: Harnessing of a river by building dams for power production.

Hydrosphere: Region that contains all of earth's water (liquid, frozen, and gas).

Hypothesis: A possible explanation of an observed phenomenon.

I

Inference: Tentative generalization based on analysis.
Integrated pest management (IPM): Use of biological, chemical, natural, and cultural controls to control a specific pest problem.
Introduced species: Species that is not native to the an ecosystem.
Ion: Charged atom.
Ionosphere: Outer layer of the atmosphere composed of ions.
Inverted pyramid: The producer base is smaller than the next higher trophic level.

J

J-shaped curve: A curve with the shape of a *J*, a population in which the size of the population does not limit its growth.

K

***K*-selected strategist:** A species that exhibits an S-shaped growth curve; lives in an ecosystem with a constant, predictable climate; has a long life-span; matures slowly; and reproduces for several years after reaching maturity.
Kilometer (km): Unit of distance equal to 6/10 of a mile.

L

Law of conservation: *See first law of thermodynamics.*
Law of entropy: *See second law of thermodynamics.*
Law of tolerance: *See limiting factors principle.*
Laws of thermodynamics: Describe the behavior of energy.
Leopold, Aldo: Founder of the land ethic concept, in which he argued that mankind has an ethical responsibility to the environment (i.e., the biosphere). Man, as part of the biosphere, has a responsibility not to abuse and mistreat the biosphere.
Limiting factors principle: Too much or too little of one or more factors will eliminate an organism from an area.
Lithosphere: Region of soil and rock making up the earth's crust, mantel, and core.
Litter layer: Undecomposed, dead, organic matter; lies on top of the A horizon.

M

Macronutrient: Element required in a large amount.
Market goods: Goods or services that are sold.

Mature soil: Gains and losses soil in equal amounts; the profile does not continue to develop but stays the same.

Maximum density: Maximum number of individuals that can eke out a living in the system without damaging its resources.

Meter (m): Unit of distance equal to one yard or three feet.

Micronutrient: Element required in a small amount.

Modern view: The idea that to understand a system you must understand its parts and how they interact; the system is the result of constant change and the interaction of its parts.

Monoculture: The planting of only one kind of crop on a piece of land.

N

Natural eutrophication: *See eutrophication.*

Natural resource: Substance used by man derived from the environment.

Natural selection: Process of selecting those individuals in the population who will live to reproduce because they, by chance, are better adapted to a selective pressure in the environment.

Natural system: System that has not been altered by human activity, is self-maintained, and is based on solar power.

Negative feedback mechanism: The product of the reaction slows down or turns off the reaction.

Neo-Darwinist: A person who believes that evolution is a slow, gradual process.

Net community productivity: The amount of energy stored at the consumer level, after the cost of cellular respiration has been subtracted.

Net primary productivity: The amount of carbohydrate fixed that is stored in excess of autotroph respiration costs; the energy that will be available to the heterotrophs in the system.

Niche: A description of a species' structural and functional role in the ecosystem.

Nitrogen-fixing bacteria: Bacteria that can take nitrogen from the atmosphere and convert it to a usable form.

No tillage: The practice of not turning over the soil to keep a continual vegetation cover on farm acreage.

Nonmarket goods: Goods and services provided free by the natural environment.

Nonrenewable resource: Substance that is produced at a rate less than the rate of consumption.

Numbers pyramid: The number of organisms at each trophic level.

Nutrient cycle: Circular pathway of an essential nutrient.

Nutrient soup: Bodies of warm water in which abiotic synthesis could occur.

O

Objectification: The process of turning things into objects, often as consumer goods, that allows people to separate themselves from responsibility for their harm or destruction.

Old soil: Loses material faster than it gains material; nutrients are leached away faster than they can be replaced.

Omnivore: An organism that obtains its energy by consuming both producers and other heterotrophs.

Opportunistic species: *See r-selected strategist.*

Optimum density: The number of individuals at a lower level of density than the maximum density.

Organism: An individual living unit.

Overdraft: Depletion of a water resource faster than it is replenished.

Oxidizing atmosphere: Atmosphere that contains free molecular oxygen.

Ozone: Molecule composed of three atoms of oxygen.

Ozone hole: Depletion, or thinning, in the ozone layer.

Ozone layer: Gaseous layer of the upper atmosphere that protects earth from the sun's high-energy ultraviolet radiation, located in the lower level of the stratosphere.

P

Panspermia: A theory that life was created somewhere else, then deposited on Earth.

Parent material: The underlying rock; partially modified parent material forms the C horizon of the soil profile.

Permeability: Characteristic of soil; indicates how fast water can move though soil.

Perpetual resource: Substance found in the environment in such large quantities that it is essentially inexhaustible.

Pheromone: A chemical that insects emit to attract potential mates.

Photochemical smog: Mixture of air pollutants produced when hydrocarbons and nitrous oxides interact with sunlight.

Photosynthesis: Process of converting radiant energy into chemical bond energy that is stored in organic compounds.

Pigment: A substance that absorbs wavelengths of light energy; it determines the color of an object.

Pioneer species: The first species to move into an area that has no developed community.

Pioneer stage: The first stage in community development.

Point-to-point-graph: A graphing technique used to present data when the information is continuous.

Population: A group of individuals, all of the same species, living in the same area.

Qualitative information: The study of how the size of a population changes over time.

Positive feedback mechanism: The product of a reaction keeps the reaction going.

Practical science: *See applied science.*

Prebiont: Nonliving, highly organized, stable droplets with characteristics found in Bionts.

Precautionary principle: The idea that before any idea or technology is used in the biosphere, it must be tested to prove that it will not harm the biosphere; safeguards must be built into any plan or technology before it can be used.

Predator: *See carnivore.*

Primary carnivore: Eats herbivores.

Primary consumer (C_1): Organism on the second trophic level; they consume herbivores.

Primary productivity: The productivity of the producers in a system.

Primary succession: Community development that occurs on sterile sites where no previous community has existed.

Producer: Autotroph that is self-nourishing; (P) organisms on the first trophic level.

Productivity: Amount of organic matter fixed by the process of photosynthesis in a given area over a period.

Punctuated equilibrium: The idea that evolution occurs in two phases: rapid change and equilibrium.

Pure science: *See basic science.*

Q

Qualitative information: Information that does not lend itself to numerical expression.

Quality reproduction: The production of few offspring; energy is expended to raise them to reproductive maturity.

Quantitative information: Information that can be expressed numerically.

Quantity reproduction: The production of a large numbers of offspring; little or no energy is expended to raise them to reproductive maturity.

R

r-selected strategist: A species that exhibits a J-shaped growth curve, lives in an ecosystem with unstable and variable climes, has a short life-span, matures early, and reproduces once.

Radiation: Propagation of energy in the form of fast-moving particle or waves.

Recycling: To put materials back into use, either by reclaiming the material and using it again or by reusing the product in the same or a different form.

Reducing atmosphere: First stable earth atmosphere without free molecular oxygen.

Reflect: To bounce off a surface.

Renewable resource: Substance produced at a rate greater than or equal to the rate of consumption.

Reserve: The total amount of a resource that can be used in a cost-effective manner.

Reservoir pool: The part of the nutrient cycle that contains a large amount of an element; the element moves out of the pool slowly; the element is either in a form that is chemically unusable or physically remote.

Residence time: How long an element stays in a compartment of a cycle.

Rich: A highly productive ecosystem that stores energy at a rapid rate.

Right side up pyramid: The producer-base level is larger then the next higher trophic level.

S

S-shaped curve: A curve with the shape of an *S;* a population in which the size of the population levels off as the population increases.

Safe density: *See optimum density.*

Salinization: Increasing the salt concentration in topsoils by overirrigation; the salt concentration is high enough to poison plants.

Saprovore: Organism that obtains its energy from decaying organic matter.

Science: People seeking to discover facts so they can gain a better understanding of the relationships in the natural and physical world.

Scientific method: The problem-solving process used by scientists.

Scientists: People with a special interest in understanding how the natural and physical world works.

Second law of thermodynamics: No spontaneous transformation of energy is 100% efficient; some energy will be changed into a dispersed form of energy that is not usable by the system.

Secondary carnivore: Eats primary consumers.

Secondary consumer (C_2): Organism on the third trophic level.

Secondary productivity: *See net community productivity.*

Secondary succession: Community development in an area in which a previous community existed.

Sedimentary cycle: A nutrient that has its reservoir pool located in the earth's crust.

Selective cutting: Harvesting only good lumber trees in a forest.

Selective pressure: Environmental factor that will influence or affect a population.

Seral stage: A transition phase in the development of a community.

Soil: The surface layer of the earth's crust; the product of the weathering of the parent material, the decomposition of organic matter, and the activities of the biotic community.

Soil profile: A cross-section of the horizons, or the sequence of layers; an expression of the physical structure of the soil.

Soil water: *See capillary water.*

Solid waste: Regularly collected waste material from households, institutions, agriculture, industry, and commercial establishments.

Specialist species: Species that have narrow ranges of tolerance.

Species: A group of individuals who are capable of interbreeding successfully under natural conditions and producing offspring that can reproduce.

Standing crop: The amount of living organic matter in a system at one point in time.

Stratosphere: The atmospheric level above the troposphere; it extends up 30 miles.

Subsistence density: *See maximum density.*

Subsoil: The accumulated material leached from the A horizon; forms the B horizon of the soil profile.

Succession: *See community development.*

Sustainable yield: The management of a resource so that the extraction rate equals the rate of natural production.

Synergism: The combined effect of two or more factors that could not be caused by each factor acting alone.

T

Taproot system: Plant has a single major root with a few smaller branch roots; it goes deep into the soil.

Teleological hypothesis: The explanation of an observation that implies that the organism knows, has a reason, or has a goal in mind, and therefore, does something to achieve it.

Terracing: Cutting a slope to create a series of horizontal steps that are at right angles to the slope of the land.

Tertiary carnivore: Eats herbivores and other carnivores.

Tertiary consumer (C_3): Organisms on the fourth trophic level.

Theory of gradual aggregation: A theory that life arose abiotically on earth over a long period of time as inorganic material formed into organic groups.

Theory of evolution: Darwin's theory that explains how life changes over time by the process of natural selection.

Theory of spontaneous generation: A theory that states that, given the correct conditions, organisms spontaneously arise from nonliving matter.

Thermal flux: Temperature changes of any surface or object above 0° C.

Top carnivore: The last carnivore in the food chain; an organism at the highest trophic level.

Topsoil: Soil that is nutrient rich; it is composed of fine organic matter mixed with gravel, sand, silt and clay; it forms the A horizon of the soil profile.

Toxic materials: *See toxic substances.*

Toxic substances: Substances that can adversely affect individuals and systems; natural or man-made.

Trace element: Elements required in a very small amount.

Transmit: To pass through.

Transpiration: The process by which autotrophs move water from the soil up to their leaves for photosynthesis.

Tree farm: An area planted so that all of the trees are of the same species and same age; a monoculture of trees.

Tree plantation: *See tree farm.*

Trophic: Energy.

Trophic classification: The classification of organisms based on where they get their energy in the chain.

Trophic level: Position in the food chain determined by the number of energy-transfer steps to that level.

Troposphere: The lower level of the atmosphere, which extends approximately seven miles up from earth's surface.

Turnover time: The time required to replace the total amount of an element in a nutrient pool.

U

Uniformatarianism: Lyell's theory that explains how the earth changes; the same slow processes that shaped the earth now were the same forces or processes that shaped it in the geologic past.

Urban sprawl: The expansion of urban areas at the expense of other natural ecosystems.

V

Value: What something is "worth."

Variation: The differences between two organisms.

W

Weathering: The break-down of parent material because of the actions of wind, rain, and temperature changes, producing pieces of different sizes that vary from gravel to clay.

Y

Young soil: Accumulates organic matter faster than it is lost; continues to develop a profile.

BOOKS AND OTHER SOURCES

The following is a selected group of books, videos, and reference information that I have found interesting, and which you might find enjoyable and informative. The material is listed alphabetically, not in order of importance or personal preference.

Books

Abbey, Edward. (1968). *Desert solitaire*. New York: Ballantine Books. (or any other books of his naturalist essays and the philosophy of environmental activism)

Adams, Douglas, & Carwardine, Mark. (1990). *Last chance to see*. New York: Harmony Books. (endangered species)

Carson, Rachel. (1961, 1991). *The sea around us*. New York: Oxford University Press. (warning of environmental ocean problems)

Carson, Rachel. (1962). *Silent spring*. Boston: Houghton Mifflin. (a warning of pesticide impact on the environment)

Craven, Margaret. (1973). *I heard the owl call my name*. New York: Dell. (story of recognition of personal relationship to the environment)

Darwin, Charles. (1957). *Voyage of the beagle*. New York: Dutton. (Darwin's own account of his voyage)

Dominguez, Joe, & Robin, Vicki. (1992). *Your money or your life*. New York: Viking Penguin. (finances and how you live your life)

Eiseley, Loren. (1957). The slit. In Loren Eiseley, *The immense journey*. New York: Vintage Press. (essay on man and geologic time)

Eiseley, Loren. (1975). The dancer in the ring. In Loren Eiseley, *All the strange hours*. New York: Charles Scribner's Sons. (essay on Darwin as a synthesizer of information and how science works)

Eiseley, Loren. (1978). The star thrower. In Loren Eiseley, *The star thrower*. New York: Times Books. (or any other of his books: naturalist essays on man's relationship to nature)

Gould, Stephen Jay. (1981). *The mismeasure of man*. New York: W.W. Norton & Co. (a reevaluation of scientific research into IQ)

Hoff, Benjamin. (1982). *The Tao of Pooh*. New York: Viking Press. (the philosophy of Taoism using Pooh and friends as examples)

Hoff, Benjamin. (1992). *The Te of Piglet*. New York: Dutton. (the philosophy of Taoism using Piglet as an example)

Hubbard, Ruth. (1990). *The politics of women's biology*. New Brunswick, NJ: Rutgers University Press. (another view of science, feminism in science)

Hubbard, Ruth, & Wald, Elijah. (1993). *Exploding the gene myth: How genetic information is produced and manipulated by scientists, physicians, employers, insurance companies, educators, and law enforcers*. Boston: Beacon Press. (critic of modern genetics and its manipulation and how it affects people's lives)

Janovy, John. (1992). *Vermilion sea*. New York: Houghton, Mifflin Co. (naturalist's view of the sea)

Kaza, Stephanie. (1993). *An attentive heart: A conversation with trees*. New York: Fawcett. (poetic and philosophical essays on trees)

Krebs, Charles. (1994). *Ecology* (4th ed.). New York: Harper-Collins. (ecology textbook)

Le Guin, Ursula. (1968). *A wizard of earthsea*. New York: Bantam Spectra Books. (fantasy fiction on the importance of naming things)

Leopold, Aldo. (1966). *A Sand county almanac.* New York: Random House. (naturalist essays on environmental ethics)

Levy, Walter, & Hallowell, Christopher. (1994). Green perspectives. New York: Harper Collins College Pub. (environmental essays)

Lindbergh, Anne Morrow. (1955). A gift from the sea. New York: Random House. (a philosophical look at the different phases of life)

Lopez, Barry. (1986). Arctic dreams. New York: Bantam Books. (naturalist essays about the arctic environment, people, plants, and animals)

Martin, A. C., Zim, H. S., & Nelson, A. L. (1951). American wildlife and plants: A guide to wildlife food habits. New York: Dover Publishing. (which plants are needed for food by different animals: useful for planning your environment)

Mayr, Ernst. (1976). Evolution and the diversity of life. Cambridge, MA: Belknap Press of Harvard University Press. (essays in evolutionary biology)

Miller, G.T., Jr. (1994). Living in the environment (8th ed.). Belmont, CA: Wadsworth Publishing Co. (environmental science textbook with case studies)

Morgan, Marlo. (1994). Mutant message downunder. New York: Harper-Collins. (story of modern and aboriginal man and how they relate to the environment)

Odum, Eugene. (1971). Fundamentals of ecology (3rd ed.). New York: W.B. Saunders, Harcourt Brace & Co. (ecology textbook)

Paley, Vivian Gussin. (1992). You can't say you can't play. Cambridge, MA: Harvard University Press. (relationships between individuals)

Saint-Exupery, Antoine De. (1940). Wind, sand and stars. New York: Harcourt Brace and Co. (philosophical and poetic story of human existence)

Saur, K.G. (1994). Encyclopedia of world problems and human potential (4th ed.). New York: K.G. Saur, R.R. Bowker Co. (environmental issues and human abilities)

Scheff, Victor. (1969). The year of the whale. New York: Charles Scribner's Sons. (a naturalist's version of the life of a whale)

Starr, Cecie, & Taggart, R. (1995). Biology: The unity and diversity of life (7th ed.). New York: Wadsworth Publishing Co. (biology textbook)

Stein, Sara. (1993). Noah's garden: Restoring the ecology of our own backyards. New York: Houghton Mifflin Co. (gardening from the environmental and ecology point of view: how to make your own balanced ecosystem)

Walker, Brian. (1994). Hua Hu Ching: The unknown teachings of Lao Tzu (rev. ed.). New York: Harper Collins. (a rendition of Lao Tzu's oral teachings of Taoism, a philosophy of man's relationship to the world)

Walker, Melissa. (1994). Reading the environment. New York: W.W. Norton & Co. (essays on the environment)

Wilson, Edward. (1992). The diversity of life. New York: W.W. Norton & Co. (issues of biodiversity and the environment)

Xerces Society, Smithsonian Institution. (1990). Butterfly gardening: Creating summer magic in your garden. San Francisco, CA: Sierra Club Books. (overview of butterfly and moth life cycles; complete resource section on butterfly gardening and plans for gardens)

Videos

Crowley, C. (Producer). (1991). In these ancient trees. The Windstar Foundation. The National Wildlife Federation, Dept. AFV, 1400 6th St. Washington, DC 20036. (issue of deforestation in the northwestern U.S.)

Griesinger, P.R. (Producer, Director). (1991). An introduction to ecological economics. Griesinger Films, 7300 Mill Rd., Gates Mills, OH 44040; (216)423-1601. (panel discussion of world ecological economics)

Griesinger, P.R. (Producer, Director). (1993). Conversation for a sustainable society. Griesinger Films, 7300 Mill Rd., Gates Mills, OH 44040; (216)423-1601. (panel discussion of how a sustainable society would work)

Griesinger, P.R. (Producer, Director). (1993). Investing in natural capital. Griesinger Films, 7300 Mill Rd., Gates Mills, OH 44040; (216)423-1601. (panel discussion of the world's natural; resources and sustainable yield)

Miller, R. (Producer), & Ballard, C. (Director). (1983). Never cry wolf. Walt Disney Productions, Anaheim, CA. (balanced look at man's relationship to the wolf)

Ramsay, P. (Producer), & Ramsay, T.E. (Director). America's wetlands. TCR Productions. Educational Video Network, Inc., 1401 19th St., Huntsville, TX 77340; (409)295-5767. (wetland ecology and future)

Russell, G.H. (Producer), & Edmondson, G. (Director). (1994). Understanding ecosystems. Educational Video Network, Inc., 1401 19th St., Huntsville, TX 77340; (409)295-57676. (good overview of how ecosystems work and environmental issues)

Sattin, R. (Producer), & Skee, M. (Director). (1990). After the warming: Parts I and II. Ambrose Videos Publishing, Inc., 1290 Avenue of the Americas, Suite 2245, NY 10104. (greenhouse effect past and future and global warming)

Temple, E. (Producer, Director). (1993). Edward Abbey: A voice in the wilderness. Eric Temple Productions, Box 2284, South Burlington, VT 05407-2284; (800)644-4747. (life of the founder of environmental activism)

Thomas, K. (Producer), & Bergstrom, K. (Director). (1991). Spaceship Earth: Our global environment. Worldlink. (today's critical environmental issues)

Waak, P. (Producer), & Diamond, J. (Director). Populations. National Audubon Society, 2020 Superior St., Sandusky, OH 44870. (the effect of world population on the biosphere)

Winslow, S. (Producer, Director). (1993). *The power of water.* National Geographic Society, PO Box 2895, Washington DC 20077-9960; (800)638-4077. (the hydrological cycle and its disruptions)

Reference Sources for Current or Updated Information

Adbusters: Journal of the mental environment. The Media Foundation 1243 W. 7th Ave., Vancouver, B.C., V6H 1B7, Canada; (604)736-9401 (an alternative look at advertisements)

Allen, John L. (Ed.). (1993). *Environment 93/94.* (12th ed. & annual editions) Guilford, CT: The Dushkin Publishing Group, Inc. (yearly updates on environmental issues)

American Chemical Society, Dept. of Government Relations and Science Policy; 1155 16th St. NW, Washington, DC 20036; (202)872-4600. (acid rain, global climate change, pesticides, waste, and water issues)

American Nuclear Society, 555 N. Kensington Ave., La Grange Park, IL 60525; (708)352-6611. (nuclear issues)

Audubon Magazine. National Audubon Society, 2020 Superior St., Sandusky, OH 44870. (bird information)

California Energy Commission. (Jan. 9, 1995). *A guide to alternative fuel vehicles.* Sacramento, CA: California Energy Commission. (information on non-gasoline driven vehicles)

Call To Action . . . , Suite #818, PO Box 917729, Longwood, FL 32791. (current legislative environmental bills and state problems)

Committee on the Conduct of Science. (1989). *On being a scientist.* Washington DC: National Academic Press. (discussion of what it means to be a scientist)

Co-op America, 1612 K St. NW, Suite 600, Washington, DC 20006; (202)872-5307. (information for direct economic action for the environment)

Earthwatch, 600 Mt. Auburn St., Box 403, Watertown, MA 02272. (information on environmental issues and projects that you can become involved in around the world)

The Earthwork Group. (1989). *Fifty simple things you can do to save the earth.* New York: Harper Collins. (booklet that describes things that you can do in your daily life to save the earth)

The Earthwork Group. (1991). *A student environmental action guide.* New York: Harper Collins. (list of things that you can do as a student to change the earth)

Electronic Mail (e-mail)
    (environmental information through your computer: issues and data available)
 • America Online
 • Association for Progressive Communications (APC)
 • CompuServe
 • Green Disk
 • Internet
 • Poptel/GeoNet
 • Right-To-Know Computer Network (RTK): access TRI
 • Toxic Release Inventory (TRI): monitoring
 • Toxnet

GOA (U.S. General Accounting Office) (1994, December). *Electric vehicles: Likely consequences of U.S. and other nations' programs and policies.* Report to the chairman, Committee on Science, Space, and Technology, House of Representatives. (update on electric vehicles) PO Box 6015, Gaithersburg, MD 20884-6015. To place an order, (202)512-6000; to obtain a daily list, (301)413-0006. (government reports and testimony)

Goldfarb, Theodore D. (1993; yearly). *Taking sides: Clashing views on controversial environmental issues.* Guilford, CT: The Dushkin Publishing Group, Inc. (pros and cons on environmental issues)

Greenpeace, 1436 U St. NW, Washington, DC 20009.
    (updates on environmental issues and how you can become involved)

Lies of Our Times, 145 West 4th St., New York, NY 10012. (environmental information)

National Geographic Magazine. National Geographic Society, PO Box 2895, Washington DC 20077-9960; (800)638-4077. (in-depth articles on human cultures and the natural world)

National Green Pages. Co-op America, 1612 K St. NW, Suite 600, Washington, DC 20006; (202)872-5307. (annual guide to green business products and services)

National Wildlife Federation, 1412 16th St. NW, Washington, DC 20036-2266; (202)797-6800. (air issues)

Nature Conservancy, 1815 North Lynn St., Arlington, VA 22209; (703)841-5300. (environmental group that conserves habitats)

New Jersey Audubon Society, 790 Ewing Ave., Franklin Lakes, NJ 07417; (201)891-1211. (state information on bird and other environmental issues)

New Jersey Environmental Federation, 94 Church St., 2nd fl., New Brunswick, NJ 08901; (908)846-4224. (environmental information for NJ)

The New York Times. (Sunday ed. & Tuesday *Science* section). (weekly update of science that effects you)

The North American Butterfly Association, 39 Highland Ave., Chappaqua, NY 10514. (all aspects of butterflying)

The real goods hard corps. Real Goods, 966 Mazzoni St., Ukiah, CA 95482-3471. (catalog with environmental products, books on alternative energy and housing)

Rocky Mountain Institute, 1739 Snowmass Creek Rd., Snowmass, CO 81654; (303)927-3851. (information on alternative energy lifestyles)

Sierra Club, 730 Polk St., San Francisco, CA 94109. (environmental issues)

Soil and Water Conservation Society, 7515 NE Ankeny Rd., Ankeny, IA 50021-9764; (800)THE-SOIL. (soil and water issues)

Swedish Nuclear Fuel and Waste Management Co., Box 5864, S-102 48 Stockholm, Sweden. (radioactive waste)

United States Environmental Protection Agency, 401 M St. SW, Washington, DC 20460; (202)260-7373. (water, acid rain, and recycling issues)

U.S. Fish and Wildlife Service, Arlington Square Building, Room 130, Dept. of the Interior, 18th and C Sts. NW, Washington, DC 20240. (wildlife issues)

U.S. General Accounting Office *See GOA.*

Water: The power, promise and turmoil of North America's fresh water. (1993, November; Special issue). *National Geographic.* (water issues)

World Watch Magazine. Worldwatch Institute, 1776 Massachusetts Ave. NW, Washington, DC 20036-1904; (202)452-1999. (current update of environmental issues)

Worldwatch Institute. *The state of the world 1994-1995.* New York: W.W. Norton & Co. (annual update of current world environmental issues and what needs to be done)

Worldwatch Institute. (1995). *Vital signs: The trends that are shaping our planet.* New York: W.W. Norton Co. (annual update showing global trends of environmental concern, published annually)

The World Wildlife Association, 1250 24th St. NW, Washington, DC 20037. (group interested in saving earth's wildlife)

Xerces Society, 10 SW Ash St., Portland, OR 97204; (503)222-2788. (group interested in the global conservation of invertebrates and their habitats)

INDEX

M

N

T

U

V